The Academic Library: Its Context, Its Purpose, and Its Operation

The Academic Library: Its Context, Its Purpose, and Its Operation

Editor

Jina Rowen

scitus
academics

The Academic Library: Its Context, Its Purpose, and Its Operation
Edited by **Jina Rowen**

ISBN: 978-1-68117-639-0
Library of Congress Control Number: 2016946499

SCITUS Academics LLC,
616, Corporate Way, Suite 2, 4766,
Valley Cottage, NY 10989

www.scitusacademics.com

Printed in United States of America on Acid Free Paper

Preface

The history of human civilization is a history of development, each generation learning from the experience of earlier generations, making their own achievements and contributing their work to the knowledge to be inherited by the next generation. Even a generation which is radical in its approach, changing much of what it has inherited, has its experience to contribute to the next generation. We learn about the achievements and the failures of the past from our parents, from our teachers, from the structures of the society in which we live. In that learning process, a process which goes on throughout our life, libraries can have a very important role. Academic library's main function is to support their parent organization in achieving the organizational objectives. As the name suggest academic libraries are involved in providing support for research and educational activities in universities, institutions, and other academic organizations. It involves content development, acquisition, technical services, providing institutional repositories, inter-library loan, and document delivery services etc. some academic libraries also establish writing centers and centers for information literacy too. The library is not just a repository, or a service like any other, or a place for study: it is all these things. It can also be a partner in research and in teaching, and institutions which fail to capitalize fully on this asset will find it harder to compete in the future. Library Operations staff are responsible for managing building hours, coordinating major and minor renovation projects, operating mail and delivery services and providing safe and secure access to library buildings and collections.

The Academic Library: Its Context, Its Purpose, and Its Operation explores exciting programs and initiatives that can both engage practitioners with academic libraries and assist academic librarians in creating a vibrant library atmosphere.

Table of Contents

Chapter 1	New Roles for Academic Libraries in Scientific Information	1
Chapter 2	Library Value in the Classroom: Assessing Student Learning Outcomes from Instruction and Collections	13
Chapter 3	Ethnography in Action: Active Learning in Academic Library Outreach to Middle School Students	39
Chapter 4	Faculty and Librarians' Partnership: Designing a New Framework to Develop Information Fluent Future Doctors	59
Chapter 5	Today and in Perpetuity: A Canadian Consortial Strategy for Owning and Hosting Ebooks	113
Chapter 6	Libraries Atwitter: Trends in Academic Library Tweeting	135
Chapter 7	The Renaissance in Library Buildings	157
Chapter 8	Acquiring E-Books for academic libraries	169
Chapter 9	The e-Reader — an Educational or an Entertainment Tool? e-Readers in an Academic Setting	177
Chapter 10	Knowledge and Skills for the Digital ERA Academic Library	191

Chapter 11 Factors that Increase the Probability of a Successful Academic Library Job Search 217

Chapter 12 How Much is Much?: A Conceptual Study of Web Traffic 239

Index 261

1

CHAPTER

New Roles for Academic Libraries in Scientific Information

Pål A. Bertnes

Faculty of Law Library Karl Johans gate 47 0162 Oslo, Norway

In this presentation I will focus on the situation and development of academic libraries both in a macro and micro perspective. The macro perspective comprises society as a whole, higher education and research – areas in which libraries believe they have little or no influence.

The micro perspective comprises conditions within the libraries or within the political and economic framework of the libraries.

My remarks are formed by 27 years of experience from the law library1 of the University of Oslo2 Judging by the number of library visitors, the number of visits to our home-pages and the size of our collections, our library is a fairly large law library by international standards. It has twenty full-time staff, out of which eighteen professional librarians, and an equivalent number of part-time students.

Society: Some Characteristics

- •It is characteristic of most European countries that knowledge, education and new forms of communication are economic values and hence important areas of commitment.

- A consequence of today's accelerating technological development is a demand for new knowledge, new disciplines and professions.
- The educational market is less national than before. The European Market and the internationalization in most fields require knowledge of other countries' culture and language.
- The increasing international competition in the areas of higher education and research affects private as well as public educational institutions.

HIGHER EDUCATION: THE UNIVERSITIES

My reflections regarding the problems and challenges of the universities are also based on knowledge of my parent organization, The University of Oslo. It has always been a public university with all funds covered by the national budget. But the trend we have seen over the last years is a reduction in the university budget and an increasing gap between the needs of the university and the allocated resources. Therefore the university must also look for external funding and income in order to increase or even maintain the quality of research and teaching. Irrespective of public or private funds, high quality is the most important factor in the competition for academic staff, students and economic resourses.

However, recruiting highly qualified academic staff by offering good salaries has not been possible at the University of Oslo and when the salary is low, it affects the social status. What remain as recruitment factors are idealism and a quest for knowledge.

The university must adapt to the new requirements of society, must meet the needs of the private and public sector and must accept the challenges of lifelong learning, remote teaching, post-graduate education, problem-based learning, interactive teaching methods and digital course-material. The physical presence of student or teacher is no longer self-evident.

I will not attempt to suggest possible strategic solutions for the university, but it is evident that it must function as a flexible and open system, a system that is willing to change - a condition for survival in today's competitive market oriented society. But the traditional values and ideas that the university has built up over the centuries must also be respected and protected in the future.

THE ACADEMIC LIBRARIES

Introduction

The academic libraries have always played an important role in the life of the universities. They have been regarded as essential to research and academic studies. Physically, they have often been situated in the middle of campus and thought of as „the heart of the university". They have also been a valuable resource for other parts of society. We see that technology today makes it possible to implant artificial hearts and manipulate the human genes. We may well wonder what implants will replace the library?

Information 328 Which Commitments Should the Academic Libraries Make?

It is said that knowledge workers will be the most important profession in this century. Who are they? In my opinion there is no doubt that they are the information specialists – our librarians. Through the years they have developed a knowledge and methodology in collection management, quality assessment and retrieval that are indispensable whether we are dealing with printed or digital information.

Even though students, teachers, researchers move in a digital world with fast access to the core body of knowledge from their personal computer, they still visit the library. Why? Because the library is committed to solve their information needs and personal communication is still the best way. The library is also a forum for discussion and a social meeting place.

This complex situation gives us, in my view, one main commitment: to serve the hybrid or multifunctional library in years to come. New Commitments – Dare We Choose?

Do we have the competence, the resources, the support to attack new areas of development? Do we know the real information needs of our users in years to come?

It will be years before we can answer these questions. Do we have time to wait for the answers? The answer is no! We must dare make choices now.

Evidently, the choice of commitment varies between libraries, depending on the institution's infra-structure and special needs. In the following I will present some of the areas chosen by our library that should be of interest to other academic libraries.

Scanning of Printed Texts

Our goal is to improve availability to material difficult to access or in other words, to meet the needs of our users in a speedy and rational way.

In our library we have had an increasing demand for translated versions of Norwegian legislation, not least from abroad. The Norwegian authorities have never approached this problem in a systematic way. Acts have been translated more or less ad hoc, many stored in the Foreign Office. We set out to collect whatever existed of translated acts and have constructed a database available on Web called Translated Norwegian Legislation3 , and we have scanned the documents into the database. Search keys are date, number, words in title (any language). Due to copyright rules not all acts are in digital form. The database also has an index of the printed acts only available in the library. In addition there are links to the official Norwegian texts in Lovdata (host for the Norwegian legal information system).

No one else had endeavoured to undertake anything of this kind before, as it had not been considered commercially interesting. To us, it has been economically interesting. We are spared the work of retrieving, copying and dispatching. The database itself was developed in co-operation between library staff and staff committed to the digitization project of the University library. There is free access. Perhaps we ought to turn it into a fee-based service? At any rate we have acquired experience in what it takes to construct a database tailored to the need of our users and we have made our work visible to the surroundings. The database is visited by lawyers all over the world.

AN ELECTRONIC BOOK

For many years we have given courses in library use and retrieval of legal sources for students at different levels. In 1997 one of my colleagues and I published the second edition of a textbook on the study of law (Praktisk rettskildelære). The Faculty of law introduced a new course in practical methodology for graduate students (Profesjonsstudiet) in which our book became part of the curriculum.

When the book was published we negociated a deal with the publisher, giving us the liberty to produce an electronic Internet version of the book. To accomplish this we managed to get external funding and technological expertise from outside. The electronic book was completed in six months.

The book has become an interactive netbook with hyperlinks to the text and to electronic sources elsewhere on the Internet, features impossible to implement in a printed version. The book is continuously updated.4

The project has given us experience on many levels and has been noticed by the faculty. The academics have, moreover, discovered a potential for their own future textbook production.

Gateways to Subject Information on Internet

Studies of the excellence of search engines show that robots are not a reliable agent for retrieving scholarly information. Out of a billion web-pages only 5 - 7 % represent academic material, the best engines merely cover 10 to 20 % and a great deal is out-dated. This chaotic picture needed some structure in order to serve our users.

For several years we collected legal links and organized them primarely according to nation. But as more and more legal information world wide became digital, Web grew rapidly in importance as a resource for the legal world and we realized that something drastic had to be done. In the summer 1999 we chose the database programme Roads as administrative tool for our link collection. First of all we had to learn the programme, set up specifications and working routines. Major goals were to organize the links according to subject and to involve most of the staff in the process. Our ambition was, and still is, that this Internet-based information service should have the same quality as our traditional paper-based collections. We have allocated 15 % of the staff to this task. Two librarians work half-time with input of data, coding, describing and assuring quality. Most of the other staff members spend about 10 % of their time supervising the Internet within their respective fields and suggesting new links to be included (title and description)

In January 2000 the two other Norwegian university law libraries in Bergen and Tromsø joined the venture and thus what started out as a collection of links turned into the Norwegian gateway to law information.5

What are the consequences for our library of this new service? In relation to new technology we have taken on a new role. This is essentially what we always have done: evaluating, describing and making available. With the gateway to law we have provided researchers and students with a good tool towards finding legal sources on the net, and we have given the reluctant ones a temptation to get started.

As there is free access to the database, the legal profession outside academia has discovered the value of our work. We are now negociating with the Norwegian Bar Association regarding co-operation in the future development of the database.

There are no great costs involved in such a project, but it entails, as all new projects, a change in priorities and the rationalization of other tasks. With a professional staff full of enthusiasm and highly IT-qualified this change has been possible.

COMPETENCE DEVELOPMENT

A lot of what we do today demands another competence than that we acquired at Library school a few years back. In today's situation it is evident that staff development/staff training is time consuming, but nevertheless essential to the organization. Extensive use of own staff for competence development strengthens competence and increases the library's general teaching qualifications. Strategies for staff development may take place on different levels:

Firstly, to ensure a general high standard of competence for all staff. Secondly, to tailor competence for special tasks (areas of work) or special groups of staff to make staff more focused on the relationship between competence and their field of work.

Personally, I am a very slow IT-learner – might have something to do with age? But that's not the only reason why I am impressed by my staff's ability to tackle new technology. They seem to have the mentality: We will figure this out! We do not want to rely entirely on the computer experts. This is one of the reasons why we have been able to engage in and succeed in so many projects. The combination of librarianship and IT knowledge gives greater insight in potential areas of development than mere computer knowledge. Therefore I believe in developing relevant IT expertise in my own organization. I must admit, however, that we sometimes need purely technological expertise and we have recently engaged a computer technician as well.

Paradoxically, having a staff qualified in IT technology represents a danger. They can be tempted to take better-paid jobs outside the academic world. In order to keep the best there must be a change in personnel policy at the university, allowing to a much larger extent than today, the possibility, for instance, of salary differentiation. But to achieve this, we need acceptance from the head of the university and understanding from the unions and professional organizations.

OTHER PROJECTS

As time is limited I will only list a few of the other projects we have undertaken, some of which are consortia based:

- Electronic journals
- Outsourcing and centralized solutions
- Research documentation and mediation
- Fundraising

Library Management

The context for library management for the last thirty years has been one of change, which means that the library constantly has to deal with new territory, adapt to new demands and refashion services to meet changing needs. For library managers there is a great challenge in inspiring staff to address all these new areas of work and thus turn the library into the flexible organization it needs to be.

Considerable time should be spent in helping staff to find ways of rationalizing everyday routines and making change in priorities. In my organization we have developed a relatively „flat“ structure which entails that all staff must know something about everything in addition to being experts within limited fields. This makes it easier to involve more people in projectrelated work and have others take over other tasks that need to be taken care of. We all know the benefits of teamwork: there is the effect of synergy and the good feeling of participating towards a common goal. For the team spirit it is important that as many as possible have ownership in the results of our achievements.

On several occasions I have emphasized the importance of promoting our services. In the struggle for resources and positions we have to improve our marketing strategies. On many occasions the head of the library could well be an important promoter. But as we address a highly skilled and intellectual „consumer“-group accustomed to making their own evaluations, the best promotion is the product itself: quality service that satisfies the users' information needs.

With networks and ressource-sharing co-operation on different levels is more relevant than ever, e.g.:

- Co-operation with the academic environment of the parent organization
- Co-operation with end-users to ensure a user-centered library
- Co-operation with other libraries, publishers and computer scientists

CONCLUDING REMARKS

Electronics has not yet superseded the paper-based environment, the printed media, on the contrary. Due to the latest in computers, network technology, digital printing and scanning methods there are more printed pages produced than ever before. But the academic libraries are only able to purchase a small fraction of the total academic production. The future trend for book acquisition will surely follow the development we have seen for periodicals: only acquisition of the publications most in demand, basic literature and the predicted future classics.

For the rest, information fulfillment will be based on a combination of electronic interlending, some kind of „last minute" delivery and retrieval of digital documents.

But this is a development that will not diminish the librarians' position as intermediaries and managers of knowledge. The challenge for the library is therefore to strengthen its position as an institution that meets the information needs of its end-users regardless of the relative importance of its own collections.

In what way have the projects I described earlier in my talk affected our library? First of all, we have created some indispensible tools to help solve our problems, but more important, we have demonstated the almost „unmeasurable", that our work has provided value-added services. This has in turn resulted in the willingness of the faculty and external funds to give more backing, to allocate more resources to our library

This spring, we have observed and noted the turbulence on the stock exchanges all over the world due to the rise and fall of computer shares. We experience, on the contrary, that our library's shares steadily rise. Why? First of all, because we are not a shareholder company or a purely commercial enterprise in a speculative branch. Secondly, because we represent an institution that works towards ideal long-term goals without a bias to short-term rewards. We value knowledge and quality above dividends and options. Therefore we must seize the opportunity to tell our environment that it pays to rely on us.

If we apply technology in a reasoned way, we will contribute to the further development and innovative use of existing tools, to the benefit of our users. Moreover, if we are open to new developments in research, learning and technology, we will always keep ahead in what is our field: academic and professional information.

REFERENCES

1. <http://www.ub.uio.no/ujur/>
2. <http://www.uio.no/>
3. <http://www.ub.uio.no/ujur/ulov/>
4. <http://www.ub.uio.no/ujur/prjus/>
5. <http://www

2

CHAPTER

Library Value in the Classroom: Assessing Student Learning Outcomes from Instruction and Collections

Denise Pan, Ignacio J. Ferrer-Vinent , Margret Bruehl

Technical Services, University of Colorado Denver, Auraria Library, 1100 Lawrence Street, Denver, CO 80204, USA

University of Colorado Denver, Auraria Library, 1100 Lawrence Street, Denver, CO 80204, USA

University of Colorado Denver, Department of Chemistry, Campus Box 194, P.O. Box 173364, Denver, CO 80217-3364, USA

ABSTRACT

What is the value of library services and resources in the college classroom? How do library instruction and collections contribute to academic teaching and learning outcomes? A chemistry instructor, instruction librarian, and technical services librarian collaborated to answer these questions by combining chemistry education and information literacy pedagogy to assess student learning. The authors developed curriculum units that teach information literacy skills and scientific literature research in a General Chemistry Laboratory course for Honors students. Their study extends beyond examining library instruction and collections assessment in isolation. Rather, their research protocol intends to contribute to student learning outcomes assessment research. The authors propose that an embedded, mixed-methodology, and longitudinal approach can be used to collect data and assess outcomes in terms that describe and measure the value of library services and resources.

Keywords: Value of library services and resources; Assessment; Information literacy; Student learning outcomes; Library instruction; Collections assessment

INTRODUCTION

The institutional goal of research universities should be a balanced system in which each scholar – faculty member or student – learns in a campus environment that nurtures exploration and creativity on the part of every member (Boyer Commission on Educating Undergraduates in the Research University, 1998, p. 10).

This story began with a simple email and lead to an interdisciplinary study conducted over three academic years. At the University of Colorado (CU) Denver, in the summer of 2010, a technical services librarian contacted a chemistry instructor and asked: "In the current economic climate, when university administrators are looking for ways to balance the budget, it is imperative that libraries provide evidence of value and demonstrate their contribution to university priorities. Are you available to discuss your potential participation in a study?" By fall term, the survey participant became a co-researcher and with the help of an instruction librarian, the library was embedded into two chemistry classes. Together the three like-minded faculty members from different areas of academia and librarianship collaborated to produce curriculum units for Honors students in General Chemistry Laboratory I and II courses, administered over fall and spring semesters. The authors exposed first-year students to scientific literature and assigned information literacy activities that help build problem-solving and critical thinking skills to engage and promote student success. Moreover, they developed and implemented a research protocol that enabled them to gather and analyze data on student learning outcomes over time.

At the conclusion of a three-year study, the three faculty members asked themselves three questions: Can we generalize this methodology? Will it scale? Does it contribute to the organizational goals of student retention and success? Their answer was, "we believe so." This article is an invitation to practicing librarians and Library and Information Science researchers to implement the CU Denver research protocol for gathering and analyzing data to measure the value of library services and resources. In this article, the authors will explain how their article contributes to the growing body of literature focused on student learning outcomes assessment; describe their research protocol and curriculum units; and provide a summary of study results. Companion articles address the case study methodology, implementation, and student performance assessments (Ferrer-Vinent, Bruehl, Pan, & Jones, submitted for publication); and describe the curriculum units developed and their connection to building information literacy (Bruehl, Pan, & Ferrer-Vinent, submitted for publication).

STUDENT LEARNING OUTCOMES ASSESSMENT

Like a rally call to the troops, the Association of College and Research Libraries (ACRL) and Megan Oakleaf created *The value of academic libraries: A comprehensive research review and report* (2010). In essence, the report encourages campus level conversation on assessment, accountability, and value. Within the context of institutional mission and outcomes, they identify "Student Success" as one of the top ten areas of library value on which to focus a research agenda. In response to the question – "How does the library contribute to student learning?" – Oakleaf states that the current literature on information literacy is "voluminous," but a majority is "sporadic, disconnected, and reveals limited snapshots of the impact of academic libraries on learning." Instead, she recommends that "Academic librarians require systematic, coherent, and connected evidence to establish the role of libraries in student learning" (ACRL, 2010, p. 118). In her review and analysis of the literature, Oakleaf introduces several practical suggestions (p. 37–42). The authors distill these concepts into three words that describe the essence of their foray into student learning outcomes assessment: collaboration, purposefulness, and longevity.

COLLABORATION

Comprehensive and meaningful assessment of student success is impossible in isolation. A learning ecosystem can be cultivated between student and instructor; student and librarian; and instructor and librarian. Poetically described by the Boyer Commission report, *Reinventing undergraduate education*, "The ecology of the university depends on a deep and abiding understanding that inquiry, investigation, and discovery are the heart of the enterprise, whether in funded research projects or in undergraduate classrooms or graduate apprenticeships. Everyone at a university should be a discoverer, a learner" (Boyer Commission on Educating Undergraduates in the Research University, 1998, p. 9). Throughout library literature, these sentiments have been echoed for more than a decade.

When evaluating the 21st century library, Smith (2001) describes the changing environment of higher education. The concept of learning has shifted from the "teacher's knowledge to the student's understanding and capabilities...it requires the faculty to bring the strength of the research paradigm into the learning process" (2001, p. 29). Academic faculty members are being asked to become learning experts by focusing on outcomes assessment — developing individual students' competencies and demonstrating collective programmatic success. As part of the academic community, the mission of the library must change from "a content view (books, subject knowledge) to a competency view (what students will be able to do)" (p. 32). No longer gatekeepers to materials or tools, academic librarians must take a more active role in the learning process and contribute student learning outcomes for academic programs across the curriculum.

Similarly, Nimon recognizes the expanded role of librarians in measuring the outcomes of academic programs. To do so, she encourages developing partnerships between the library and academic departments to teach information literacy. Moreover, the success is contingent on including assessment criteria that reflect the goals of all stakeholders — librarian, academics, and students. She explains, "Student evaluation of the program must be appropriately tailored to show whether its goals were readily visible to the learners and whether the learners considered them met...It will be necessary for the assessment of student work to be at least in part a joint responsibility" (Nimon, 2001, p. 50).

Participation in pedagogy and assessment activities is not just the role of academic librarians, but their obligation. Bundy asserts that they "can no longer responsibly disengage from why students want the print and digital information and resources to which libraries can now so readily provide access. Nor can they disengage from whether those students have the capacity to apply that information well, and to what use they put it" (2004, p. 2). Instead, both academic teachers and librarians should be immersed in the total educational process — including program and curriculum development, learning design, pedagogies, assessment, and the scholarship of teaching and learning.

Information literacy is an important concept that should be owned by all educators (Bundy, 2004, p. 7). However, this terminology may not be recognizable outside the library walls. Some examples of synonyms and overlapping concepts that may resonate more widely include the following: information, research, or 21st century skills; independent scholarship or research; lifelong learning; scientific method; research processes; and Bloom's taxonomy. As Oakleaf explains, "For those facing greater challenges, establishing and using a common language that emphasizes shared campuswide values may produce greater success" (2011, p. 65).

In his model for academic libraries, Lewis (2007) proposes a strategy for maintaining the central library position on campus in the digital age. One of his top five strategies is to "Reposition library and information tools, resources, and expertise so it is embedded into the teaching, learning, and research enterprise...Emphasis should be placed on external, not library-centered, structures and systems" (2007, p. 3). By focusing on student learning, academic libraries and librarians have new opportunities to reestablish their place on campus, engage with their colleagues, and maximize their contribution to their institutions and higher education as a whole.

PURPOSEFULNESS

When developing student learning outcomes assessments, academic librarians, and faculty members should proceed with purpose. In other words, they are intentionally gathering and analyzing particular types of data. Specifically, Dugan & Hernon (2002)assert, "student learning outcomes are concerned with attributes and abilities, both cognitive and affective, which reflect how the student experiences at the institution supported their development as individuals" (2002, p. 377). In addition, "outcomes assessment alerts us to what students know or do not know about research," Carter explains, "thus allowing librarians to adapt instruction to the needs of the students" (2002, p. 41). This requires a commitment to document, evaluate, and communicate impacts on student learning, as well as improving one's own teaching and assessment skills (Oakleaf, 2011, p. 70).

The literature on information literacy assessment identifies numerous techniques and tools. These tactics require a range of resources or experiences and assess various perspectives. Radcliff et al. describe three different learning domains: *affective* assesses how students feel or their opinions; *behavioral* evaluates what students can do; and*cognitive* measures what students know. They classify performance assessments, such as report writing, as part of the behavioral domain (2007, p. 19–20, 115). In contrast Oakleaf describes performance assessments as "real-life applications of knowledge and skills" and "[they] reinforce the concept that what students learn in class should be usable outside the classroom" (2008, p. 239). All can agree that assignments with an information literacy component can be used to measure higher-order thinking skills, and to achieve greater integration and contextualization in an academic course. While the results can offer a high degree of validity, they may have limited generalizability (Radcliff et al., 2007 and Oakleaf, 2008).

The literature on assessment of information literacy is quite bountiful — well documented in monographs, handbooks, manuals, guides, and articles. The focus is nearly exclusively on library instruction, and overlooks other areas of librarianship. Library collections are included in the broader context of library assessment and emphasize cost effectiveness (Hufford, 2013, p. 20–26). Kinman's article "E-metrics and library assessment in action" is a rare example that highlights the significant role electronic resources play in demonstrating the value of libraries and impacting on student learning outcomes (2009). There is ample opportunity to expand this area of the scholarship into a new line of inquiry that captures the complexity of the learning environment and inspires more rigorous and critical investigation.

LONGEVITY

More than library instruction, information literacy extends outside the confines of the library and extends to the classroom, campus, and beyond. Each level requires more collaboration and complex logistics (Iannuzzi, 1999, p. 304). To be truly meaningful, outcomes assessment is an ongoing process. It requires time, effort, resources, and collaboration with academic faculty. The research process can be simple or elaborate, but must be based on meaningful data. From each experience, researchers can learn, refine, and improve assessment procedures (Carter, 2002, p. 41). Oakleaf & Kaske (2009) recommend that librarians follow best practices, use multiple methods and strategies, adjust goals and objectives, and repeat assessments continuously over time (2009, p. 283).

CU Denver research protocol & curriculum units

The literature describes collaboration, purposefulness, and longevity as key ingredients to achieve student learning outcomes assessment. This article continues this conversation. The authors at CU Denver adopted these assessment perspectives and adapted them to their local circumstances. Specifically the authors summarize their student learning outcomes assessment research protocol in three components: collaborating internally and externally; articulating outcomes; and assessing over time. By doing so they are able to expand beyond their silos and approach student learning outcomes assessment from an interdisciplinary perspective. The results of their study demonstrate the rich data and insights that are possible from their approach.

Collaborating internally and externally

The CU Denver researchers identified multiple opportunities for collaborations within librarianship, academia, and the campus. They found commonality with academic faculty and librarians from diverse areas and expertise. In addition, they also created teaching, research, and learning opportunities between faculty and students. A visual representation of these connections is available in Fig. 1.

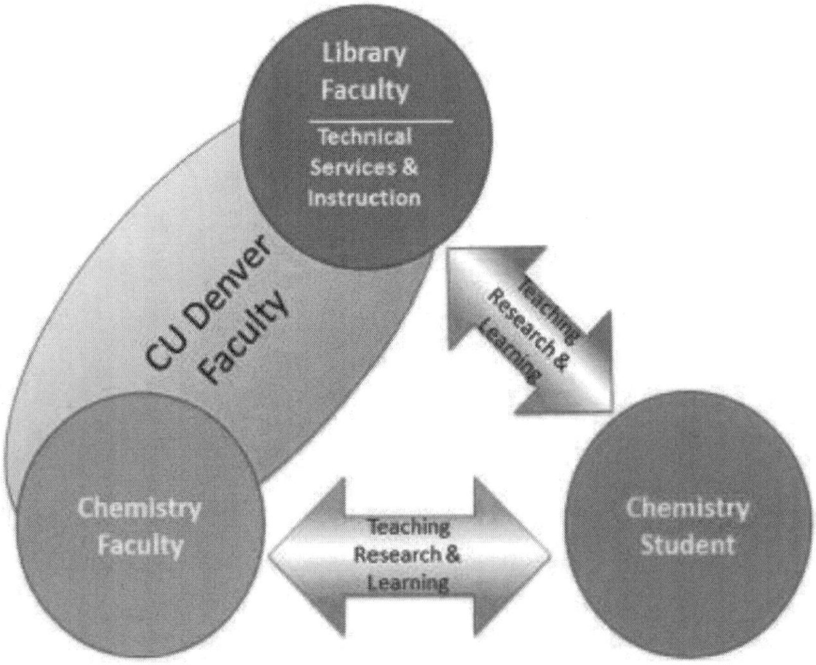

Figure 1. Research relationships.

At CU Denver, librarians have faculty status. The library and chemistry departments are two separately administered academic units reporting to the Office of the Provost. Although the departments often cooperate on collections and instruction initiatives, the authors' research study represents a new level of collaboration. Library instruction and collections assessment activities are embedded into two Chemistry courses. More specifically, the researchers focused on shared values and appreciated other perspectives from different disciplines. In essence, they are proponents of a "golden triangle" approach to curriculum design "where everyone contributed to each component with the outcome being more robust than if they used the more linear model" (Fox & Doherty, 2012, p. 151–2).

Within the library context, a technical services librarian and an instruction librarian from the same academic library developed a partnership to conduct a student learning outcomes assessment research study with a chemistry instructor. This experience may be atypical. Bundy states that "few university libraries or librarians directly engage with, or reach out to, other parts of the profession...the focus tends to be on information resource sharing and access, rather than on learning collaborations and strategies" (Bundy, 2004, p. 9).

Most significantly, the CU Denver faculty developed a research relationship with their students. With Colorado Multiple Institutional Review Board approvals, they were granted permission to gather data from the students. In her Post Laboratory Assignment, Bruehl specifically invited her students to participate in the study and assured them that no additional work would be assigned, "Simply completing this assignment is all that is required." All personal identification was removed by Bruehl before Pan and Ferrer-Vinent analyzed the student data. Bruehl also explicitly told the students that they could contact her during lab or via email to option out of the study and their data would be excluded from the study (Bruehl et al., submitted for publication).

Articulating outcomes

Sharing common values and goals, the authors found exciting synergies for developing student learning outcomes assessment. All three researchers aspired to develop a project that could provide compelling evidence of supporting the institution's mission for campus administrators. They surmised that an assignment that concentrated on information literacy could help them achieve this objective. In doing so, they developed a research instrument that enabled them to gather data on students' ability to find, access, and evaluate scientific literature.

When developing her Honors General Chemistry I and II Laboratory courses, Bruehl aims to create a student centered learning environment. This teaching philosophy encourages active participation by students while the learning process is facilitated by the instructor. The investigative and inquiry-based laboratory courses emphasize teaching students problem-solving and critical thinking skills through open-ended experiments and using specialized techniques and instrumentation. This teaching and learning strategy builds on the CREATE (Consider, Read, Elucidate hypothesis, Analyze and interpret data, Think of the next Experiment) method.

Typically, the real language and research process documented in primary scientific literature is not presented to students until their upper division course or not at all during their undergraduate education. This delay is recognized as a missed opportunity at best, or a loss of science majors at worst. Using the CREATE pedagogy exposes students to scientific research literature and demonstrates the creative and exploratory nature of collecting and interpreting data for research. By doing so, students are being taught how to think like scientists, increase engagement, and could help with retention (Hoskins et al., 2007 and Gottesman and Hoskins, 2013). Findings from one study indicate that the CREATE method "increases students' confidence in their ability to read and understand primary literature, improves their self-assessed understanding of the nature and processes of science, and encourages their development of more sophisticated epistemological beliefs" (Hoskins, Lopatto, & Stevens, 2011, p. 375).

The CU Denver research study has two overarching goals. The first is to demonstrate that teaching information literacy skills in first-year chemistry courses can provide immediate and long-term benefits to student performance. The second is to quantify the benefits students received in their educational activities by using library collections. Long term, the authors hope to be able to provide campus administrators with evidence that their efforts contribute to meaningful outcomes, such as student retention, graduation rates, and economic benefit.

To achieve these goals, the authors developed two complimentary curriculum modules to establish a foundation of information literacy skills for beginning science students using scientific literature and laboratory experiments. The first unit, administered in the fall semester, is entitled "*Introduction to Scientific Literature*" and is comprised of three components: 1) formal library instruction by a science librarian; 2) reading and in-class discussion on a scientific journal article; and 3) pre- and post-lab exercises to research an idea for a new general chemistry experiment. The students are asked to develop information literacy skills by searching the scientific literature for concepts, recording their research process, refining their premise, writing a brief description of their proposed experiment, and citing three resources in American Chemical Society (ACS) format that directly support their proposal.

Building on the skills and information they learned in the fall, in the spring semester they are asked to conduct a multi-week project entitled "*Design your own General Chemistry Lab.*" Again students explore the scientific literature for ideas on which to base their experiment, and record the reference for any resources they downloaded and read. Their final project includes a formal proposal; documented laboratory procedure that has been designed, developed, and tested by the student; citation list in ACS format; and presentation of results to the class.

In both semesters, students are conducting research in the scientific literature. To record their research process, Bruehl requires students to complete a research process template. The format of the template is a table with three separate columns to record: 1) database/search tool name; 2) search terms/refinements; and 3) resources viewed. The third column instructs the students to "paste ACS or specific journal format citation here."Table 1 includes an example of a completed research process template, with good searching and narrowing values, provided by a student in spring 2012.

Table 1. Sample research process template.

Database/search tool name	Search terms/refinements	Resources viewed (paste ACS or specific journal format citation here)
Journal of Chemical Education	Simple and titration and experiment	Lopez, E.; Vassos, B.H., Colorimetric titration experiment for the undergraduate laboratory. *J. Chem. Educ.* **1984**, *61(11)*, 1025.
Journal of Chemical Education	Titration and content of	Deal, S.T.; Pope, S.T., Improved end point detection in the redox titration of vitamin C in green peppers. *J. Chem. Educ.* **1996**, *73(6)*, 547.
Journal of Chemical Education	Titration and vitamin content	Sowa, S.; Kondo, A.E., Sailing on the "C": A vitamin titration with a twist. *J. Chem. Educ.* **2003**, *80(5)*, 550.
Journal of Chemical Education	Titration lab and vitamin C	Virz, E., A student laboratory experiment based on the vitamin C clock reaction. *J. Chem. Educ.* **2007**, *84(7)*, 1156.
Journal of Chemical Education	Titration lab and vitamin C	Conn, L.W.; Johnson, A.H., Vitamin C. Content of frozen orange and grapefruit juices. *J. Chem. Educ.* **1933**, *25(2)*, 218–221.
ScienceDirect	Titration and vitamin C	Suntornsuk, L.; Gritsanapun, W.; Nilkamhank, S.; Paochom, A. Quantitation of vitamin C content in herbal juice using direct titration. *J. Pharm. Biomed. Anal.* **2002**, *28(5)*, 849–855.
ScienceDirect	Titration and vitamin C	Lenghor, N.; Jakmunee, J.; Vilen, M.; Sara, R.; Christian, G.D.; Grudpan, K. Sequential injection redox or acid–base titration for determination of ascorbic acid or acetic acid. *Talanta.* **2002**, *58(6)*, 1139–1144.
Google	Vitamin C titration equations	Bacon, J.R. Determination of vitamin C by an iodometric titration. *Western Carolina Univ.* **2006**.

The research process template is an essential assessment tool for the authors to gather data to evaluate the students' ability to find, access, and evaluate scientific literature. Bruehl required the student to record *all* resources viewed, even if the student only read the abstract or the table of contents, and not the article itself. She explains that, "Some resources that you record in this template will be more important to your project. These resources help to formulate your design, and they must ALSO be listed in your formal citation list" (Bruehl et al., submitted for publication).

By documenting the names of database/search tool, search terms, and resources viewed, the authors could see how students are selecting the resources viewed — rejecting or accepting results that fall in and out of the scope, or distract or meet the needs of their topic. Broad subjects are narrowed with keywords, and then filtered down by publications focused on science experiments, such as the *Journal of Chemical Education*. Combining the expertise from academic faculty and faculty librarians, the authors analyze and compare the data from the research process templates and citation lists. Bundy summarizes this partnership and librarian contributions to teaching pedagogy as "Disaggregated roles, such as assessing learning resources for quality, overlap with what librarians do now, and the subject expertise of the academic teacher is being married with the librarian's navigation and sense making of the information universe" (Bundy, 2004, p. 9).

Assessing over time

Realizing that outcomes assessment is a continuous process, the CU Denver researchers intentionally developed the project to be repeatable over multiple semesters and academic years. The process encourages collaborative teaching and learning from the students and one another. The authors refine and improve instructional strategies and assessment methods at each encounter. For example, during citation analysis Ferrer-Vinent and Pan observed that a few students were citing articles that were not available full-text from the library. While it is possible that the students could have requested the article from interlibrary loan, the librarians were worried that these students did not understand the difference between an abstract and the full-text article. They shared their concerns with Bruehl who added this information to her lecture on scientific literature.

Exposing students to information literacy concepts over multiple semesters reinforces skills and the interconnectedness of information. The importance of the research skills and process is reiterated over two terms of an academic year. In the spring semester *"Design your own General Chemistry Lab"* assignment, Bruehl reminds students to expand their understanding beyond their own experience, and to use the research expertise they developed the previous semester. Scientific literature offers "a vast array of resources" to assist them with their assignment, which in turn helps them to develop and recognize sound scientific procedures, evaluate experimental methods, and draw appropriate conclusions. As a result, beginning chemistry students are fine tuning their ability to use the scientific method to investigate a scientific question.

Moreover the CU Denver study was conducted over three academic years. The researchers analyzed consistencies and variations over the same and consecutive semesters across multiple academic years. For example they could examine data for each term (e.g. Fall 2010 and Spring 2011), or comparison between terms (e.g. Fall 2010 versus Spring 2011), and within an academic year (e.g. 2010–2011). Most importantly, since the chemistry students agreed to participate in the research project, the researchers surveyed the same students 1–2 years after they finished the Honors General Chemistry II Laboratory course. Therefore, with the CU Denver research protocol, the researchers gathered quantitative and qualitative data from the students at different points of their academic careers — during their first-year and beyond.

CU Denver research study results

On their own, the researchers could have independently assessed outcomes from their own disciplines' perspectives. By collaborating internally and externally, articulating outcomes, and assessing over time, the authors developed and implemented their research protocol from fall 2010 to spring 2013. Expanding beyond their silos and using an interdisciplinary perspective and mixed methodology approach enabled them to produce distinct student learning outcomes assessment data and analysis.

Prior to their partnership, each author individually engaged in assessment activities. Bruehl conducted student learning outcomes assessment by grading students' assignments, exams, and overall performance in the class. Ferrer-Vinent provided in-class library instruction, surveyed student perceptions, and assessed students' post-instruction performance. With no direct contact with students, Pan measured the library's potential contributions by analyzing usage statistics of electronic resources.

Together they leveraged personal expertise to enhance their shared research project. Their interdisciplinary approach allowed them to develop and evaluate richer data sets. The CU Denver researchers apply the quantitative citation analysis data from the research process template and qualitative surveys to address their research inquiries. Specifically, they intend to demonstrate that teaching information literacy skills in first-year chemistry courses can provide immediate and long-term benefits to student performance; and to quantify the benefits students received in their educational activities by using library collections. Research outcomes are summarized in three subsequent sections: 1) quantitative information literacy study results; 2) quantitative collections ROI study results; and 3) qualitative student benefits survey results. As previously mentioned, details on the authors' research methodology is described in their companion case study article (Ferrer-Vinent et al., submitted for publication).

Quantitative information literacy study results

By recording the references of resources viewed and cited in their assignment, the authors could begin to quantify and aggregate how students are developing information literacy skills. For example, Table 2 demonstrates that they were able to calculate the average number of resources viewed per student to complete the fall and spring term assignments and their average course grade. For the fall semester assignment, students were instructed to use their library skills to view resources and then select three upon which to base a proposed experiment. In spring, students received no guidance on how many resources should be viewed or selected to support their experimental design. Excluding the first semester, during which the authors were still refining the assignment guidelines, students on average looked at almost 10 resources in the course of their assignments and their average grade was 87.5%.

Table 2. Resources viewed and final course grade.

	# of students	Total # resources viewed	Min # resources viewed	Max # resources viewed	Average # resources viewed	Average final grade %
AY2011						
Fall 2010	26	174	3	16	6.7	89.8%
Spring 2011	21	197	1	28	9.4	90.7%
AY2012						
Fall 2011	37	369	3	39	10.0	84.5%
Spring 2012	36	341	0	66	9.5	87.1%
AY2013						
Fall 2012	25	248	5	26	9.9	87.9%
Spring 2013	16	155	4	13	9.7	87.1%
Total	161	1484	0	66	9.2	87.5%

The researchers intentionally collected grades to enable them to correlate academic performance and use of library resources. Said another way, they hypothesize that there is a relationship between the extensive use of library resources and a high score. If students consulted multiple databases, reviewed many articles, and selected the best three resources, they should receive a higher score than a student who only used one database and looked at a few articles. The authors collaborated with Galin Jones – Associate Professor at the University of Minnesota, School of Statistics – to determine that there is a statistically significant positive relationship between resources viewed and the students' final grade for the course. The full explanation on the linear regression analysis is available in the case study companion article (Ferrer-Vinent et al., submitted for publication).

Quantitative collections ROI study results

The authors also use the citation analysis data to calculate the cost benefits of purchasing scientific literature to support learning objectives. To do so, they measure library value by deriving value using return on investment (ROI). According to Tenopir (2012), "Derived values, such as return on investment (ROI), use multiple types of data collected on both the returns (benefits) and the library and user costs (investment) to explain value in monetary terms" (2012, p. 6). In order to measure return on investment, the authors needed to assign a monetary value to having access to non-market resources/services or library collections. They applied contingent valuation as defined by Megan Oakleaf to assign value to library collections and to determine potential willingness to pay to maintain the existence of library collections (ACRL, 2010, p. 50). Many publishers, including ACS, sell individual articles online. If the library did not provide these articles, students could purchase these ACS articles on their own. The authors utilize the price to download an article to derive a market value for the library service of providing access to collections. In turn, this calculated value enabled the researchers to calculate student benefits for the ROI calculations.

The CU Denver Student ROI Model is based on established ROI and cost benefit analysis (CBA) formulas. ROI is calculated as a percentage. It shows the return or increase in value on dollars spent to achieve a benefit. The generic formula is benefits minus costs divided by costs and multiplied by 100.

$$((\text{BENEFITS}-\text{COSTS}) \div \text{COSTS}) \times 100 = \text{ROI}.$$

CBA uses the same values as ROI. However, CBA is the ratio showing the dollar value of benefits gained for dollar value of costs. The basic formula is benefits divided by costs.

$$\text{Benefits} \div \text{Costs} = \text{CBA}.$$

Benefits are the estimated cost for students to buy cited articles directly from the publisher with pay-per-view. Costs are the Library costs to supply cited online journals. This methodology is based on a parallel CU Faculty ROI Model and multi-campus study Pan conducted to measure the institutional value of library resources used by faculty in their research (Pan et al., 2013).

The authors analyzed the citations and determined that students viewed 1489 resources over three years. Most of these articles were searched on the ACS Journal Database 58.0% of the time (863 out of 1489). Collectively the students looked at 1153 journal articles from 200 unique journal titles, and they consulted the *Journal of Chemical Education* (*JCE*) 66.7% of the time (769 out of 1153). Since a majority of the students used the *JCE*, the use and cost estimates for this journal title was used to calculate the ROI and CBA for the course. The authors derived the student costs from the number of times *JCE* was viewed, multiplied by the pay-per-viewed cost. The library cost for *JCE*was estimated by taking the database cost divided by the number of titles in the database.

Since ROI is calculated as a percentage, values over 100% are a positive ROI. Similarly, since CBA is a ratio showing how much is gained for every dollar spent, $1.00 is the breaking point. Values greater than $1.00 are a "positive" CBA. The results of the CU Student ROI Model for Honors General Chemistry I and II Laboratory courses at CU Denver from fall 2010 to spring 2013 are summarized in Table 3. Overall, the results indicate a very strong ROI and CBA for the *Journal of Chemical Education* with the greatest ROI of 2324% and CBA of $24.24 in fall 2011.

Table 3. Students ROI study results for *Journal of Chemical Education.*

	ROI	CBA
AY 2011		
Fall 2010	626%	$7.26
Spring 2011	884%	$9.84
AY 2012		
Fall 2011	2324%	$24.24
Spring 2012	1109%	$12.09
AY 2013		
Fall 2012	1971%	$20.71
Spring 2013	288%	$3.88[a]
Total	1144%	$12.44

[a] The researchers attribute the low ROI of 288% and CBA of $3.88 in spring 2013 to low student enrollment due to a scheduling conflict and cancellation of one section of the course.

Qualitative student benefits survey results

In addition to collaborating on quantitative citation and cost benefit analysis, the researchers also conducted qualitative assessments. After the students completed the Honors General Chemistry II Laboratory course, Bruehl emailed them an anonymous survey. This was an opportunity to gather data on the impact of teaching information literacy to first-year chemistry students over the long term. The questionnaire reminded the students that "as part of your lab work, you used the [scientific] literature and captured your on-line searches in a research process template," and invited them to participate in the survey as a "follow up to those activities." A copy of the longitudinal student survey is available in the companion article focused on the curriculum units (Bruehl et al., submitted for publication).

The responses to the survey were overwhelmingly positive. Over 40% (37 out of 88) of the students completed the questionnaire. When asked about their experiences before and after Honors General Chemistry Laboratory, more than half stated that they searched a scientific literature database since course completion. In addition, this was nearly a 40% increase over their experience prior to taking the course (see Table 4).

Table 4. Reponses to follow up survey.

	# of students surveyed	# of student responses	# of responses that searched database *prior* to class[a]	% of response *prior*	# of responses that searched database *prior* class[b]	% of responses *since*	% change increase
AY 2011	23	8	4	50%	8	100%	50.0%
AY 2012	40	18	9	50%	17	94%	41.5%
AY 2013	25	11	1	9%	6	55%	37.9%
Total	88	37	14	38%	31	84%	38.6%

[a] Answered "yes" to the question "Prior to your enrollment in Honors General Chemistry Laboratory, had you ever searched a scientific literature database?"
[b] Answered "yes" to the question "Since you have completed Honors General Chemistry Laboratory, have you searched a scientific literature database?"

Surprisingly, students reported that they searched the American Chemical Society (ACS) journals for scientific literature more frequently than Google (77.4% and 71.0% consecutively). SciFinder, ScienceDirect, and Web of Science were also identified as scientific literature databases that they used after finishing the course. The two other resources identified were PubMed and JSTOR (more details provided in Table 5).

Table 5. Database usage since completing Honors General Chemistry Laboratory.

	# of responses that searched database *since* class	Scientific literature databases used						
		ACS	Google	ScienceDirect	SciFinder	Web of Science	Wiley	Other
AY 2011	8	6	6	4	7	5	0	0
AY 2012	17	14	13	12	9	7	3	2
AY 2013	6	4	3	2	3	3	0	0
Total #	31	24	22	18	19	15	3	2
% of Total		77.4%	71.0%	58.1%	61.3%	48.4%	9.7%	6.5%

Table 6. Explanation for database searches.

	# of responses that searched database *since* class	Explanation			
		Course	Research project/internship	Curiosity	Other
AY 2011	8	8	3	6	
AY 2012	17	13	9	9	1
AY 2013	6	7	2	1	
Total #	31	28	14	16	1
% of Total		90%	45%	52%	3%

More or less than half of the students responded that they did the search for their own curiosity or for a research project/internship (see Table 6). The vast majority of students, however, conducted these database searches for another course. The knowledge they acquired was directly applied to 17 other Chemistry courses. More significantly, their information literacy skills transferred to 21 courses in Biological Sciences, 2 in English, and 1 in Psychology. These findings support the authors' hypothesis that teaching information literacy skills can provide students with long-term benefits.

CONCLUSION

With the CU Denver student learning outcomes assessment research protocol, three researchers from different areas within librarianship, academia, and the campus collaborated to explore the value of library services and resources in the college classroom. Moreover, a chemistry instructor, instruction librarian, and technical services librarian sought to create a methodology for describing and measuring the benefits of library instruction and collections to academic teaching and learning. They discovered that their process includes three components – collaborating internally and externally, articulating outcomes, and assessing over time; and echoes the elements identified in the student learning outcomes assessment literature – collaboration, purposefulness, and longevity.

This study demonstrate the possibilities when academic faculty and librarians move beyond established roles and responsibilities, and attempt student learning outcomes assessment from an interdisciplinary perspective. The authors developed a research process template to gather data used to evaluate the students' ability to find, access, and evaluate scientific literature. In turn, this citation analysis was used to assess the efficacy of teaching information literacy skills in first-year chemistry courses and the value of purchasing scientific literature to support learning objectives. With their mixed methodology approach, the research also combined qualitative assessments with their quantitative citation and cost benefit analysis.

The CU Denver research protocol has just begun to scratch the surface by focusing on assessing information literacy outcomes within the library and the chemistry laboratory. What would happen if similar curriculum units were adapted and applied to different disciplines? Can we teach students to think like historians, sociologists, economists, etc. by exposing them to the research literature? There is ample opportunity to extend this protocol to other disciplines and beyond their institution. Iannuzzi explains that, "...if we want to ensure that those skills are applied within other courses, that there is meaningful transfer to other learning environments, and that ultimately the quality of the student's work is improved, the assessment methodology moves beyond library control into collaborative efforts with teaching faculty" (Iannuzzi, 1999, p. 304). The CU Denver experience is an example of engaging with faculty, asking for input, suggesting a partnership, and co-creating an interdisciplinary study. In essence, exciting opportunities can be possible when librarians leave the library, meet and engage with academic faculty, and capitalize on serendipity.

ACKNOWLEDGEMENT

We would like to thank the University of Colorado Denver Honors General Chemistry I and II Laboratory students who provided the data for this article.

REFERENCES

1. Association of College and Research Libraries (ACRL) (2010). Value of academic libraries: A comprehensive research review and report. Researched by Megan Oakleaf. Chicago: Association of College and Research Libraries (Retrieved from www.acrl.ala.org/value).

2. Boyer Commission on Educating Undergraduates in the Research University (1998). Reinventing undergraduate education: A blueprint for America's research universities. NY: Stoney Brook (Retrieved from ERIC database. (ED424840)).

3. Bruehl, M., Pan, D., & Ferrer-Vinent, I. J. (2014). Demystifying the Chemistry Literature: Building Information Literacy in first-year Chemistry Students through Student- Centered Learning and Experimental Design. Journal of Chemical Education (submitted for publication).

4. Bundy, A. (2004). Beyond information: The academic library as educational change agent. Paper presented at the 7th International Bielefeld Conference ((February 3–5)Germany.

Retrieved from http://conference.ub.uni-bielefeld.de/2004/proceedings/bundy_rev.pdf).

5. Carter, E.W. (2002). "Doing the best you can with what you have:" Lessons learned from outcomes assessment. The Journal of Academic Librarianship, 28(1), 36–41.

6. Dugan, R. E., & Hernon, P. (2002). Outcomes assessment: Not synonymous with inputs and outputs. The Journal of Academic Librarianship, 28(6), 376–380.

7. Ferrer-Vinent, I. J., Bruehl, M., Pan, D., & Jones, G. (2014). Introducing Scientific Literature to Honors General Chemistry Students: Teaching Information Literacy and the Nature of Research to First-Year Chemistry Students. Manuscript submitted for publication in the Journal of Chemical Education.

8. Fox, B. E., & Doherty, J. J. (2012). Design to learn, learn to design: Using backward design for information literacy instruction. Communications in Information Literacy, 5(2), 144–155.

9. Gottesman, A. J., & Hoskins, S. G. (2013). CREATE cornerstone: Introduction to scientific thinking, a new course for STEM-interested freshmen, demystifies scientific thinking through analysis of scientific literature. CBE Life Sciences Education, 12(1), 59–72.

10. Hoskins, S.G., Lopatto, D., & Stevens, L.M. (2011). The C.R.E.A.T.E. approach to primary literature shifts undergraduates' self-assessed ability to read and analyze journal articles, attitudes about science, and epistemological beliefs. CBE Life Sciences Education, 10(4), 368–378.

11. Hoskins, S. G., Stevens, L. M., & Nehm, R. H. (2007). Selective use of the primary literature transforms the classroom into a virtual laboratory. Genetics, 176(3), 1381–1389.

12. Hufford, J. R. (2013). A review of the literature on assessment in academic and research libraries, 2005 to August 2011. Portal: Libraries and the Academy, 13(1), 5–35.

13. Iannuzzi, P. (1999). We are teaching, but are they learning: Accountability, productivity, and assessment. The Journal of Academic Librarianship, 25(4), 304–305.

14. Kinman, V. (2009). E-metrics and library assessment in action. Journal of Electronic Resources Librarianship, 21(1), 15–36.

15. Lewis, D. W. (2007). A model for academic libraries 2005 to 2025. Paper presented at "Visions of Change" ((January 26, 2007), Sacramento, California. Retrieved from http://hdl.handle.net/1805/665).

16. Nimon, M. (2001). The role of academic libraries in the development of the information literate student: The interface between librarian, academic and other stakeholders. Australian Academic & Research Libraries, 32(1), 43–52.

17. Oakleaf, M. (2008). Dangers and opportunities: A conceptual map of information literacy assessment approaches. portal: Libraries and the Academy, 8(3), 233–253.

18. Oakleaf, M. (2011). Are they learning? Are we? Learning outcomes and the academic library. The Library Quarterly, 81(1), 61–82.

19. Oakleaf, M., & Kaske, N. (2009). Guiding questions for assessing information literacy in higher education. Portal: Libraries and the Academy, 9(2), 273–286.

20. Pan, D., Wiersma, G., Williams, L., & Fong, Y. (2013). More than a Number: Unexpected Benefits of ROI Analysis for

Collection Development. The Journal of Academic Librarianship, 39(6), 566–572.

21. Radcliff, C. J., Jensen, M. L., Salem, J. A., Burhanna, K. J., & Gedeon, J. A. (2007). A practical guide to information literacy assessment for academic librarians.Westport, CT: Libraries Unlimited

22. Smith, K. R. (2001). New roles and responsibilities for the university library: Advancing student learning through outcomes assessment. Journal of Library Administration, 35(4), 29–36.

23. Tenopir, C. (2012). Beyond usage: Measuring library outcomes and value. Library Management, 33(1/2), 5–13.

3

CHAPTER

Ethnography in Action: Active Learning in Academic Library Outreach to Middle School Students

Samantha Godbey, Nancy Fawley, Xan Goodman, Susan Wainscot

Education Librarian, University of Nevada, Las Vegas

Head, Library Liaisons Program, University of Nevada, Las Vegas

Health and Life Sciences Librarian, University of Nevada, Las Vegas

Science, Technology, Engineering and Mathematics Librarian, University of Nevada, Las Vegas

ABSTRACT

This article describes an outreach activity developed and coordinated by academic librarians as part of a state program for low-income middle school students. Rather than offering a traditional library tour, the library organizers wanted to provide the middle school students with a meaningful experience that would encourage active participation, critical thinking, and alleviate library anxiety. As a spin on the traditional tour, students applied an ethnographic approach to learning about the library. The authors describe the development and implementation of the activity and provide recommendations for other librarians involved in outreach to K-12 students.

Keywords: academic libraries, outreach, early recruitment, ethnography, active learning, library anxiety, STEM

Academic libraries commonly engage in outreach to their surrounding communities. The desire to adequately prepare elementary and secondary students for college education further draws universities' attention to outreach to local schools in particular, and the potential for early recruitment of young students can provide additional incentive to universities.

The University of Nevada, Las Vegas (UNLV), established in 1957, is the only public university located in the southern region of Nevada and, as such, is in a unique position to attract local residents. An overwhelming majority (84%) of UNLV undergraduates are Nevada residents. The university is also recognized as a minority-serving institution (MSI) with 49.6% of UNLV's approximately 24,000 undergraduate students identifying as a racial or ethnic minority (University of Nevada, Las Vegas, 2014)

The UNLV mission statement explicitly refers to community outreach, pledging to create a campus community that is diverse and to actively engage with the local community through outreach and research. The core theme of outreach to local residents includes support of college readiness programs that support and/or recruit local residents. A variety of outreach initiatives are in place throughout the university to support college readiness efforts, including several Gaining Early Awareness and Readiness for Undergraduate Programs (GEAR UP). A national competitive grant program of the United States Department of Education, all GEAR UP programs support cohorts of low-income students for multiple years, starting in middle school, by providing educational support and guidance to students and their families to increase students' likelihood of college enrollment and success. Participating colleges and universities also have the opportunity to tailor their activities to serve as early recruitment efforts for participating students. This article describes the challenges and successes of the UNLV Libraries' contribution to one of several GEAR UP projects in which UNLV participates, specifically, the Nevada State GEAR UP program.

The UNLV Libraries consist of one central library and three branch libraries with extensive collections and services. The libraries' mission extends beyond collections and materials and emphasizes the importance of supporting learners in their research efforts, academic success, and their pursuit of lifelong learning. As a part of the library mission and vision to support learners, librarians participate in outreach efforts aimed at local residents who may enroll at UNLV in the future. The UNLV Libraries' 2011-2015 strategic plan strengthens the aim of outreach by emphasizing the role of the university library in "contribut[ing] to the institutional mission through engagement with external communities" (UNLV Libraries, n.d., p. 4). This strategic direction highlights ways the library will engage with the external community, one of which is a K-12 Teacher-Librarian Institute for Clark County School District teacher librarians. The organization outreach mission also extends to the librarian level where the Science, Technology, Engineering and Mathematics (STEM) librarian has identified one goal of her job as assisting the University in STEM literacy outreach to the region.

The Nevada State GEAR UP program is unique in its focus on STEM careers. In preparation for campus visits, the librarians were challenged to develop an activity that was engaging and relevant to middle school students while introducing the services and support an academic library can provide for college students. Librarians developed an active-learning, inquiry-based program that addressed the unique interests of these young students, whose visit to the university was their first introduction to the independence and academic rigor of college. Over the course of the 2013-2014 academic year, 200 students from seven local middle schools visited UNLV's main library as part of the Nevada State GEAR UP program. These library visits were one component of half-day campus visits, during which students explored spaces on campus such as science facilities and an art museum and met with different campus partners such as admissions staff. Rather than offering a traditional library tour, the library organizers wanted to provide the middle school students with a meaningful experience that would encourage active participation and critical thinking. A spin on the traditional tour, loosely based on the idea of ethnographic studies, was developed (Noe, 2009). Students were given a map of the library with a few guided questions and shown two floors of the library. They were instructed to be "ethnographers" and observe how students used the library, its services, and its spaces.

GEAR UP

The GEAR UP program is a U. S. Department of Education grant program designed to increase the number of low-income students prepared to enroll and succeed in postsecondary education programs. Grant recipients, such as colleges, universities, and their partners receiving the grants work with a cohort of students from low-income schools from the seventh through twelfth grades. Low-income status is defined by enrollment in a school where 50% or more of students in that grade are either eligible for the federal free or reduced-price lunch program or by residence in public housing. GEAR UP program services are made available to all students enrolled in the school and grade level (the cohort) regardless of individual student income status (U.S. Department of Education, 2014). Funding may be provided to local or state applicants for comprehensive mentoring, outreach, and supportive services for each cohort of students. Statewide programs, such as the Nevada State GEAR UP project that the UNLV Libraries worked with, must also include a scholarship component for participating students.

The Nevada State GEAR UP program has a STEM focus and three overarching goals: to increase student and family knowledge of college options, preparation, and financing; student academic performance and college preparation; and student high school graduation and college entrance rates (Nevada GEAR UP, n.d.). Students receive in-class and extracurricular instruction, mentoring, and activities such as field trips to university, college, and community college campuses. While the Nevada State GEAR UP program has a STEM focus, students receive information about all available undergraduate programs and services. The university libraries were contacted by and our continuing involvement is coordinated by the UNLV GEAR UP College Ambassador, a staff position funded by the Nevada State GEAR UP program. The university libraries did not receive funding from the grant and participated as part of the libraries' mission and strategic plan direction to provide outreach to the local community, particularly the K-12 education community.

LITERATURE REVIEW

K-12 Outreach

Librarians, concerned with the level of preparedness they observe in college freshmen, have started to reach out to local schools to develop programs or collaborations that can better prepare students for college-level research. UNLV Libraries, for example, holds a 3-day institute during which local K-12 school librarians and classroom teachers work to develop assignments and assessments that embed developmental information literacy into their curriculum. A key element of this institute is the opportunity to interact with academic librarians and view authentic collegelevel assignments in order to better prepare K-12 students for the undergraduate experience. While many outreach programs are librarian-driven and focus on helping students develop information literacy or lifelong learning skills, the programs can also serve as part of a broader recruitment effort on the part of the university.

Much of the K-12 outreach is general in nature; however, libraries are starting to follow the lead of universities and becoming part of broader STEM outreach initiatives. Two examples serve to illustrate the possibilities of engaging in wider outreach programs. Librarians at the University of South Alabama developed a library-skills curriculum as part of a summer enrichment program for local high schools to encourage and prepare minority students to enter health care professions (Rossini, Burnham, & Wright, 2013). The Ohio State University Library system provides the collections and services for a local STEM-focused high school, which has no school library (Herring, 2013).

Academic libraries are not alone in their mission to help students develop an interest in STEM fields. College and university librarians can look to their peers in public and school libraries for further ideas on supporting STEM initiatives. School librarians can encourage students' interest in STEM fields by linking their personal experiences to science-learning environments (Subramaniam, Ahn, Fleischmann, & Druin, 2012) The authors suggest that school libraries can serve as a "hybrid space to bridge the formal classroom with the broader world" (p. 163) by linking interests in science-themed books, science programs, media technology, and social networks to STEM disciplines. Public libraries can partner with community and business groups with STEM interests for financial support, resources, and volunteers for events. Displays on modern technology and innovative inventions, plus nonfiction and science-fiction books, are additional ways to create an interest in STEM fields (Hopwood, 2012).

Library Anxiety

Another goal of the UNLV librarians in this project was to alleviate potential library anxiety among participants. Library anxiety is well-documented in the library literature, beginning with Mellon's 1986 study that developed the first grounded theory about library anxiety among university students. Research into library anxiety delves into how students perceive librarians, libraries, and the experience of library use. In Mellon's study, for example, 75% to 85% of participants mentioned fear or anxiety as they described their feelings about the library, using terms such as "scary, overpowering, lost, helpless, confused, and fear of the unknown" (Mellon, 1986, p. 162). Library anxiety studies have found that the experience of library anxiety is similar to other well-known types of anxiety such as research, math, and statistics anxiety (Onwuegbuzie, Jiao, & Bostick, 2004). Library anxiety is further described as consisting of five factors: barriers with staff, affective barriers, comfort with the library, knowledge of the library, and mechanical barriers (Onwuegbuzie, Jiao, & Bostick, 2004, p. 36). When confronted with having to conduct library research, students exhibit feelings of anxiety and not knowing how to navigate the academic environment or whom to approach for help.

Bostick (1992) found that students' high levels of anxiety relate to barriers with staff, such as an aversion to approaching library staff to ask for help. Other studies have concluded that library anxiety is often state-based, meaning that students have to be at the library or contemplating using the library to experience anxiety (Jiao & Onwuegbuzie, 1999). Students specifically express anxiety as it relates to approaching library staff and comfort with library policies and library resources.

The library user's affective response, or feeling toward use of the library facilities and library staff, has been further discussed in recent work by Project Information Literacy (Head, 2013). In this broad national study, college freshmen engaged in research described the university library collections as "ginormous" (Head, 2013, p. 12) and the process of finding and selecting resources for projects as "overwhelming" (Head, 2013, p. 11). Further, many students believed that "self-sufficiency" (Head, 2013, p. 19) was important in college and that asking for help, whether from a librarian, a professor, or tutor, would be looked down upon.

Library Tours

Library tours are a standard service, especially for first-year students. Some librarians enthusiastically lead legions of students through the library highlighting key services and spaces to students, while others believe that the time could be spent more efficiently with library instruction targeted to an assignment and given at the point of need. A 2000 survey of Association of Research Libraries member libraries focused on tour administration, structure, and librarians' perceptions of their formats. The authors found that a large majority considered traditional tours to be most effective and an important part of the academic success of the student. The guided walking tour was the most popular format. Alternative, but less frequently used formats, such as self-guided tours using printed materials or video or web formats, addressed the perceived shortcomings of the traditional tours, such as staffing requirements and ineffective use of librarians' time (Oling & Mach, 2002).

More recent articles highlight innovative approaches to library tours that, once developed, make fewer demands on a librarian's time. An augmented reality tour was developed at the University of Houston-Clearwater to answer frequently asked questions (Berrish, Jambhekar & Yue, 2013). Tours that are tied to an assignment are also deemed to be more successful because there is an impetus to complete it successfully. A MP3 tour was integrated into a graded assignment for the First-Year Information Literacy Program at University of South Carolina Upstate (Kearns, 2010). Brigham Young University used Quick Response (QR) codes, a type of two-dimensional barcode, to update the MP3 audio tours that were part of a required assignment in first-year courses (Whitchurch, 2012). While the literature suggests libraries are experimenting with alternative methods to these tours, anecdotal evidence, and the experience of the authors, indicates that the traditional, librarian-led tour is still the most popular method of introducing students to a library's space and services. However, a lack of assessment of these librarian-student interactions gives librarians no clear indication of the tours' effectiveness.

Active Learning in Libraries

Active learning is not a new concept in library instruction. Allen (1995) noted the potential positive impact of active learning on library instruction and argued that "Active learning strategies provide the opportunity to offer students the freedom and the responsibility to be themselves, to utilize their interests and learning styles in the content and design of their educational experiences" (p. 91). Jacobson and Mark (1995) discussed the potential for active learning to empower learners in the library classroom. A 2002 survey (Hollister & Coe, 2002) of over 2,000 instruction librarians revealed that those librarians favored active learning as an instructional strategy, but noted that active-learning activities can be time-consuming given the time limitations imposed on library instruction. More recently, Booth (2011) has noted that information professionals tend to have less systematic training in educational theory and pedagogy than other educators. Booth has emerged as one of the strongest proponents of a constructivist approach to library instruction and interactions and acknowledges the challenge presented to librarians in avoiding excessive detail in library instruction because they, as experts, can forget how to approach topics from the inexpert perspective of their students. She draws on work of constructivist theorists such as John Dewey to encourage librarians to "shift the focus of instruction from passive to active, from educator to learner, and from cognition to learning community" (Booth, 2011, p. 39). Further, active learning strategies have been found to have a positive effect on student perceptions of librarians and library resources, as well as decreased library anxiety (Detlor, Booker, Serenko, & Julien, 2012).

The specific active learning strategy incorporated in this project used an ethnographic method rather than a traditional library tour.

Ethnography involves "the close observation of social practices and interactions," with ethnographers "analyzing multiple types of data, such as interviews, direct observation, photographs, journals, or cultural artifacts" (Duke & Asher, 2011, p. 3). A recent analysis of the use of ethnographic methods in libraries (Khoo, Rozaklis, & Hall, 2012) noted an increase in the use of these data-collection methods in order to help improve library services and/or instruction. These ethnographic studies cover topics such as the evaluation of information literacy instruction (Pencek, Nelson, & Brians, 2009), wayfinding in libraries (Hahn & Zitron, 2011) and website usability (Klare & Hobbs, 2011). One large-scale ethnographic study is the Ethnic Research in Illinois Academic Libraries (ERIAL) Project, a two-year ethnographic study across five Illinois universities that attempted to better understand how college students conduct research and use library resources and services (Duke & Asher, 2011). Ethnographic methods as an instructional strategy is less documented, however. The group decided to follow the model set by librarians at Auburn University (Noe, 2009; Carter, 2011) and turn the middle school library visitors into ethnographers.

THE PROJECT

Development of Activity

The UNLV Libraries were contacted in September 2013 to potentially host up to 3 hours of activities for 225 students who would be visiting the campus the following month, 3 days of 75 students each. The three initial contacts in the library were Sue Wainscott, STEM Librarian; Xan Goodman, Health and Life Sciences Librarian; and Samantha Godbey, Education Librarian.

These three librarians had coordinated other STEM outreach efforts. They also reached out to Nancy Fawley, Head of the Library Liaisons Program, due to the possible magnitude of the project.

Goals proposed by the GEAR UP College Ambassador focused on increasing students' ability to succeed in rigorous academic courses. Proposed learning outcomes included educating the students about the function of the libraries in the learning process and educational mission of the university, STEM-focused resources available through the libraries, and career resources at the UNLV Libraries.

In conjunction with the GEAR UP College Ambassador, we decided that rather than hosting all 75 students for 3 hours at a time, we would conduct a one-hour workshop with groups of 25 students at a time. Students would be visiting other departments on campus in groups of 25 in order to alleviate the burden on any single campus partner. Considering the amount of time allotted to the activity, we would focus on the first of the learning outcomes proposed by the GEAR UP College Ambassador: to educate the students about the libraries' role in the learning process.

We were aware that the participants would be a mix of seventh and eighth graders, and that for most of these students, this would be their first experience in an academic library. In fact, they would most likely have a range of personal experience with libraries. We were also attentive to the fact that these students still had 5 or 6 more years in this program, and we would likely be meeting with these students multiple times in the years that would follow. Not only would we not want to repeat activities for the same students, but we would need to scaffold any instruction across potential visits over multiple years. Furthermore, since the activity would not be tied to a particular academic project or assignment, we were aware that making the activity relevant and worthwhile to the students would be a challenge.

The activity would need to be developmentally and age-appropriate. What we had done in the past with undergraduates would not work here. With college students, we generally work toward a specific assignment. Even high school students have a more immediate interest in college and are doing research that is often comparable to what can be expected in the university setting. This activity, however, would need to be situated within the middle school context. We focused on developing an active learning activity that would draw on students' prior experience and give them an introduction to the to the academic library experience.

As a group, the librarians decided to focus on introducing students to the function of the libraries in the learning process and decreasing library anxiety. Through doing so, we could make the first steps toward increasing the students' college readiness, enrollment, and success. Having identified these priorities, we realized that we didn't want the students to spend their entire library visit in a classroom; we wanted them to actually experience the library environment. One of the authors had inverted the traditional library tour at a previous institution by allowing students to take the lead. We decided that this kind of student-led exploration would be the main activity we would use with the seventh and eighth graders. We also decided to include student library workers in the activity, in order to give our visitors the opportunity to interact directly with undergraduate students.

Ethnographer for a Day

Introduction. Upon arrival, students were led to a library instruction room. Students were welcomed by one of the librarians and led in a short discussion to activate prior knowledge about libraries. Students responded to questions about their experience with their school libraries and public libraries and any differences they might have noticed between the two. Students were asked whether they had any prior experience with an academic library; these students had none. Students were also introduced to the concept of ethnographic research. We explained that rather than leading them on a tour, we would be asking them to carefully observe the library to see what they could learn themselves .

Librarians and student workers introduced themselves to attendees, including their subject areas of expertise or majors, and then dispatched groups of four to eight students out into the library. Each group included a librarian or student worker, but the GEAR UP students were encouraged to take the lead.

Observe. The students were provided with a handout on which to record their observations of the library. The handout included blank maps of two floors of the library. Service desks were labeled with stars. Students were asked to label each service desk and observe the library as they explored. The handout also included questions such as: How would you use this space? What do you observe about the library on this floor? What do you observe about the people on this floor? Finally, a question in the margin of the handout encouraged the visitors to engage with the library faculty and student workers in their groups about their college and career paths.

Students were given 20 to 30 minutes to explore the first and top floors of the library. We selected the first floor, as this floor has all of the library's service desks, and a range of study and computing spaces. The top floor includes book stacks, study spaces, and the library's most striking views – of the library itself from the top of an atrium, and of the city and Las Vegas Strip from the library windows. These two floors would give students a sense of the services and spaces that the library offers and provide them with a visually memorable experience.

The middle school visitors were very interested in observing college students eating in the library, watching nonacademic videos on the library computers (e.g., "He's watching The Walking Dead!"), and studying in groups. Students were impressed by the size of the library as compared to their school libraries, and lamented some of the stricter policies and limited hours at their own schools. Students asked detailed questions about the broader college experience, including whether college students have curfews and who does their laundry. They also asked the UNLV libraries staff and students questions about their own careers and educational experiences, including how they managed to pay for their degrees.

Share. After the allotted time, groups returned to the library classroom to report about their experience. During these final 10 minutes of the activity, students noted what they had observed, reflected on the experience, and speculated about their own future lives as college students.

RECOMMENDATIONS

Include College Students

As noted in the description of the activity, participants were placed in groups of four to eight students for the activity. We found that this size group was ideal. Larger groups can be difficult to manage and give personal attention, while students can become self-conscious in groups smaller than four. Including library student employees in the activity was extremely helpful. During the final debrief of the activity, the GEAR UP students asked a number of questions about college life that were directed toward the college students, including "Why did you pick UNLV?" Each group of middle school students was assigned to a group with a librarian or other staff person. Ideally, each group would include a librarian and a student employee.

Allow the Students to Lead

The ethnographic approach to introducing the library allowed GEAR UP students to pose questions about the library environment rather than simply taking part in a library tour. However, this approach was challenging for library faculty and staff, who were accustomed to leading activities themselves. Our student workers, who have been explicitly trained to lead library tours, found this to be especially difficult. As experts about our own libraries, we want to share our expertise, but by allowing the visitors to observe and determine the focus of their attention, those visitors can take ownership of their learning and identify aspects of the library with which they can relate. At the same time, some of the middle school students were reluctant to be seen observing and writing notes as they walked around the library. One strategy we found helpful was to pretend to address the group in strategic locations in the library. That way, the students appear to be looking at the librarian but instead can examine what was happening behind the librarian.

Integrate Assessment of Outcomes

After each visit, student participants offered feedback to the GEAR UP College Ambassador about their experience. Comments were overwhelmingly positive, and students mentioned specific librarians and students by name, suggesting that we may have been successful in reducing librarian anxiety. A recommendation for future outreach activities is to incorporate direct assessment of outcomes, rather than relying solely on anecdotal feedback. Assessments could include a survey of the students at the end of the activity or later in the academic year. Student worksheets could also be collected and scored on a rubric related to our stated outcomes.

Be Deliberate about Scaffolding

Further, it is important to consider scaffolding within the activity itself and across years with a multi-year initiative. For example, during the activity, we drew on students' prior knowledge

before introducing new concepts. We set the stage for the activity by asking students if they had experience in other types of libraries, school or public, in order to introduce the university library to middle school students. To further reduce library anxiety over time, we plan to build on this activity to introduce students to the library as a place for conducting research and library faculty as resources to assist with research. For example, we might add an activity in future years in which students explore career paths with an active learning activity using one of the databases available to all state residents.

Be Strategic about Outreach

Finally, outreach must be strategically pursued. There will doubtless be many opportunities for the university libraries to participate and partner with student recruitment projects, STEM outreach programs, and other university-led outreach activities. In fact, it is difficult at times to say no to worthy projects due to time or other constraints. Using the university libraries' strategic plan as an initial filter for opportunity selection, followed by the consideration of staff's individual goals, can be useful in evaluating potential outreach opportunities. Moreover, a deep understanding of the university's mission and strategic goals will allow us to proactively seek out additional opportunities that offer the best fit and outcomes for university libraries' outreach and service activities, with an eye toward building strategic long-term partnerships.

CONCLUSION

The Nevada State GEAR UP program is a multi-year initiative, and students will continue to make college and university campus visits. At press time, plans have not been made for the following year's visits; however, librarians are planning to develop an activity that scaffolds on the ethnographic tour of the previous year. As students progress in grade level, the focus will be on developing critical thinking skills that will prepare them for higher education and to integrate activities that draw on the library's collections.

As academic librarians, we have little contact with middle school students in a professional context. The wide-eyed surprise of students witnessing the differences between UNLV's library and their school libraries was fun to observe. The low-stakes nature of the activity allowed students to engage with librarians and library spaces in a novel way that lessened their anxiety and increased their confidence. Allowing the students to discover these differences on their own, without a traditional lecture or tour from a librarian, gave them a chance to form their own opinions about higher education and their potential future role as college students.

REFERENCES

1. Allen, E. E. (1995). Active learning and teaching: Improving postsecondary library instruction. Reference Librarian 24(51-52), 89-103.

2. Berrish, K., Jambhekar, N., & Yue, C. (2013). Augmented reality tour: Using new technology to import old information. Texas Library Journal, 89(2), 70-73.

3. Booth, C. (2011). Reflective teaching, effective learning: Instructional literacy for library educators. Chicago: American Library Association.

4. Bostick, S. L. (1992). The development and validation of the Library Anxiety Scale (Doctoral dissertation). Retrieved from ProQuest Dissertations & Theses Full Text.

5. Carter, T. (2011). Turning students into ethnographers: A library discovery tour. Presentation at Georgia Conference on Information Literacy. Retrieved from http://digitalcommons.georgiasouthern.edu/cgi/viewcontent.cgi?article=1039&context=ci l-2012-2

6. Detlor, B., Booker, L., Serenko, A., & Julien, H. (2012). Student perceptions of information literacy instruction: The importance of active learning. Education for Information 29(2), 147–161.

7. Duke, L. M., & Asher, A. D. (2011). College libraries and student culture: What we now know. Chicago: ALA Editions.

8. Hahn, J., & Zitron, L. (2011). How first-year students navigate the stacks: Implications for improving wayfinding. Reference & User Services Quarterly, 51(1), 28–35.

9. Head, A. (2013). Learning the ropes: How freshmen conduct course research once they enter college. Retrieved from http://projectinfolit.org/images/pdfs/pil_2013_freshmenstudy_fullreport.pdf

10. Herring, D. H. (2013). 21st century learning: Providing academic library services for Ohio's first STEM-focused high school. Reference Librarian, 54(4), 280-296.

11. Hollister, C. V., & Coe, J. (2003). Current trends vs. traditional models: Librarians' views on the methods of library instruction. College & Undergraduate Libraries, 10(2), 49–63.

12. Hopwood, J. (2012). Initiating STEM learning in libraries. Children & Libraries: The Journal of the Association for Library Service to Children, 10(2), 53-55.

13. Jacobson, T. E. & Mark, B.L. (1995). Teaching in the information age: Active learning techniques to empower students. Reference Librarian, 24(51-52), 105–120.

14. Jiao, Q.G., & Onwuegbuzie, A.J. (1999) Identifying library anxiety through students' learningmodality preferences. Library Quarterly, 69(2), 202-216.

15. Kearns, A. (2010). An iPod (MP3) library tour for first-year students. College & Undergraduate Libraries, 17(4), 386-397.

16. Khoo, M., Rozaklis, L., & Hall, C. (2012). A survey of the use of ethnographic methods in the study of libraries and library users. Library & Information Science Research, 34(2), 82– 91.

17. Klare, D., & Hobbs, K. (2011). Digital ethnography: Library web page redesign among digital natives. Journal of Electronic Resources Librarianship, 23(2), 97–110

18. Mellon, C. (1986). Library anxiety: A grounded theory and its development. College & Research Libraries. 47(2), 160-165

19. Nevada GEAR UP. (n.d.). Nevada state GEARUP goals and objectives. Accessed 15 December, 2014, from http://gearupnv.org/wp-content/uploads/2013/01/NEVADA-STATE-GEARUP-GOALS-AND-OBJECTIVES.docx

20. Noe, N. (2009). Ethnographers for an hour. In R. L. Sittler & D. Cook (Eds.), The Library Instruction Cookbook (p. 22). Chicago: Association of College and Research Libraries

21. Oling, L., & Mach, M. (2002). Tour trends in academic ARL libraries. College & Research Libraries, 63(1), 13-23.

22. Onwuegbuzie, A.J., Jiao, Q.G., & Bostick, S.L. (2004). Library anxiety theory, research, and applications. Lanham, MD: Scarecrow Press.

23. Pencek, B., Nelson, S. G., & Brians, C. L. (2009). What students tell us about doing research: Information literacy assessment activities as pedagogy. American Political Science Association Conference Papers, 1–39.

24. Rossini, B., Burnham, J., & Wright, A. (2013). The librarian's role in an enrichment program for high school students interested in the health professions. Medical Reference Services Quarterly, 32(1), 73.

25. Subramaniam, M. M., Ahn, J., Fleischmann, K. R., & Druin, A. (2012). Reimagining the role of school libraries in STEM education: Creating hybrid spaces for exploration. The Library Quarterly, 82(2), 161-182.

26. University of Nevada, Las Vegas, Office of Decision Support. (2014). Undergraduate student profile fall 2014. Retrieved from http://ir.unlv.edu/IAP/Reports/Content/UndergraduateStudentProfile_Fall2014.aspx

27. UNLV Libraries. (n.d.). UNLV libraries strategic plan: July 2011 through June 2015. Retrieved from http://www.library.unlv.edu/about/strategic_plan11-15.pdf

28. U.S. Department of Education (2014). Programs: Gaining early awareness and readiness for undergraduate programs (GEAR UP). Retrieved from http://www2.ed.gov/programs/gearup/index.html

29. Whitchurch, M. (2012). A quick response: QR code use at the Harold B. Lee Library. Reference Librarian, 53(4), 392-402.

4
CHAPTER

Faculty and Librarians' Partnership: Designing a New Framework to Develop Information Fluent Future Doctors

Rachid Bendriss , Reya Saliba , Sally Birch

Weill Cornell Medical College in Qatar, Qatar Foundation, Education City, Doha, P.O. Box 24144, Qatar

ABSTRACT

Upon faculty request, information literacy instruction was fully integrated into an English for Academic Purposes course for non-native English speaking students pursuing a medical degree at a U.S. institution in the Gulf State of Qatar. Adopting the flipped classroom modality, librarians designed modules to meet the students' information literacy needs while adapting the content to the course syllabus. Content was uploaded to the learning management system, Canvas. Readings, online tutorials, quizzes and assignments were created to achieve the outcomes of each session. Completion of information literacy modules is factored in the overall grades of students in the English for Academic Purposes course. This paper will outline the design, implementation, and assessment of information literacy sessions and will highlight the importance of collaboration between faculty and librarians to set a common ground and reach expected outcomes. It will also share the students' perceptions of the effectiveness of integrating information literacy in an English for Academic Purposes course.

Keywords: First-year college students; Library instruction; Information literacy; Flipped classroom; ESL

INTRODUCTION

In its framework for 21st century learning, the Partnership for 21st Century Skills (2015)established information literacy (IL) as a fundamental component of student success in the global economy. Additionally, with increased student mobility and a thriving transnational education, learning English as a Second Language (ESL) is still regarded as an essential component of an effective academic and career readiness to succeed in a knowledge-based economy. Putting these two pieces together, the present study sought to explore the efficacy of combining IL skills with an English for Academic Purposes (EAP) course for non-native English speaking first-year college students to reach a level of information fluency.

Purpose of the study

The aim of the study was to examine first-year Arab students' experience with integrating IL skills instruction in the curriculum of an EAP course at a pre-medical program of a U.S. private college in the Gulf State of Qatar.

Terminology

This study uses the terms English as a Second Language (ESL) and English for Academic Purposes (EAP) interchangeably.

Research questions

- R1: What were the IL needs of newly-graduated high school EAP Arab students enrolled in a first-year premedical program at a U.S. private college?
- R2: How effective were the IL modules in improving students' research skills as established by the Association of College and Research Libraries (ACRL)'s five standards?
- R3: What were the study participants' perceptions of the IL content as part of the EAP curriculum?

LITERATURE REVIEW

Librarians and faculty collaboration

IL skills are deep-rooted in higher education. The Middle State Commission on Higher Education (2009) describes the relationship among general education skills as inherent and intrinsic and states that "the interrelatedness is evident in the concept of "information literacy", which embraces all of the specific general education skills" (p. 47). Furthermore, Calkins and Kelley (2007) reported that students' search of the literature typically lacks rigor, and they are prone to selecting the first few references they find, regardless of quality or relevance to the topic. Moreover, undergraduates often select online sources but have difficulty assessing the credibility of the sources.

Faculty and librarians both recognize the lack of IL skills among college students.Bowles-Terry, Davis, & Holliday (2008) identify librarians and writing instructors as "longtime allies that share the goal of teaching information literacy." Through collaboration, librarians become more visible and accessible. Students can identify, relate to, and easily approach librarians for help. Dhawan and Chen (2014) believe a learning community can be created and "students form bonds" (p. 417) by linking the content of the two courses and working closely with faculty and librarians.

In many institutions, IL instruction is offered as a single stand-alone lecture. Pedagogically, this is not ideal as generally many of the tools and concepts covered in IL presentations are not used by the students until much later in the semester, increasing the likelihood that the information will be forgotten or remembered incorrectly. To overcome this challenge, McCue (2014) suggests that librarians need to create effective partnerships and build strong working relationships with faculty members responsible for the development of curricula in courses where IL instruction is present or potentially included. McCue (2014) also points out that research into IL and Blended Learning is in an early stage with the current body of knowledge consisting of case studies and small action-based research projects.

Moselen and Wang (2014) describe a systematic model for integrating IL into an undergraduate program developed by Li Wang at the University of Auckland. This model defines three key elements: the *What, Who* and *How*. While the *What* element includes the expected outcomes and the *How* element describes the method used for integration and contextualization, the *Who* element highlights the notion of collaboration between librarians and faculty as main stakeholders. The authors found that "good professional relationships with academic staff... would enable collaboration." However, some librarians participating in the study observed that a lack of confidence and a lack of knowledge of the curriculum prevented them from approaching faculty. Other librarians pointed out that faculty hesitated to ask for their help as they "did not really know what subject librarians could do to help them" (Moselen & Wang, 2014, p. 118).

Integrating IL using the flipped classroom modality

While some literature supports teaching general principles that can be applied to different contexts (Bruner, 2009), other studies found that course-integrated instructions are more beneficial as they encourage critical thinking and information evaluation rather than generic information retrieval skills (Dhawan & Chen, 2014). Getting students "to apply, synthesize, question, understand, and communicate" is described as the main challenge that librarians and faculty are facing (Bowles-Terry, Davis, & Holliday, 2008), as students usually concentrate on the assignment requirements rather than explore the topic in depth and critically evaluate sources.

In an article published in 2008 comparing e-learning to lectures, Reime et al. describe e-learning as a teaching method that uses information technology to develop, deliver and facilitate the learning process (Reime, Harris, Aksnes, & Mikkelsen, 2008, p. 799). Technology is redefining instruction, and Learning Management Systems (LMSs) are being used more to deliver e-learning in college settings. "Young students, taking an online course is not a novel experience... [They] are usually comfortable with the technology as well as a student-centered learning environment" (Jackson, 2014, p. 468).

The flipped classroom concept is being increasingly used to deliver instruction in an online environment. Youngkin (2014) describes the flipped classroom as "an inverted model of teaching that often uses various communication or instructional technologies" (p. 368). Students can do the reading, watch video tutorials and submit quizzes outside the classroom, while in-class time is used for activities, discussion and assessment. The flipped classroom emphasizes the responsibility of individual learners in reviewing and absorbing teaching material on their own rather than relying on the instructor to explain the content. It creates "a more personalized academic experience" (Youngkin, 2014, p. 369). Dhawan and Chen (2014) see the flipped classroom as an opportunity for librarians "to teach IL with some depth" while allowing students to have more time to process IL instruction. Assessment can be built-in for short-term as well as long-term goals. Quizzes, assignments, discussions and surveys can be designed and delivered using a LMS.

Arnold-Garza (2014) describes how librarians used the flipped-classroom model to successfully teach course-integrated IL sessions at Albert S. Cook Library at Towson University. Pre-library session assignments included guides, brief video tutorials, and a quiz to test students' understanding of the material. Among the topics covered were scholarly and popular sources, database search, navigating the library website, and other information related tasks. In class, hands-on activities were administered "to apply the concepts introduced by the pre-library session assignment" (Arnold-Garza, 2014, p. 11). Based on the survey results, Arnold-Garza (2014) reported that 90% of 148 students found the online materials helpful and completed the pre-library meeting assignments prior to coming to class. The authors further reported that 86% of students commented that the face-to-face meetings were important to solidify their acquisition of the literacy skills presented via the flipped modality.

Utilizing the flipped modality along with the integration of IL into an established academic course is a new concept for literacy skills instruction and has limited representation in available literature.

First year ESL college students

In the United States, many institutions are striving towards improving first year college transition as it is considered very important for students to acquire needed skills such as time management, effective study habits, research skills and critical thinking for a successful academic experience. Dhawan and Chen (2014) asserted that first year college is an important part of a national conversation in higher education. In her article entitled "Empowering ESL students," Conteh-Morgan (2001) believes IL and ESL have "overlapping theories, objectives, and practices" (p. 31). She further explains that by teaming up with faculty, librarians are given the chance to align IL instruction with the course content and make its integration into an ESL course or program an easy task.

Many authors describe the integration of research skills into a specific course as the most effective method to successfully teach IL skills (Bowles-Terry et al., 2010 and LaGuardia, 2011). It is crucial for students to put into practice these skills in a defined context to achieve specific learning outcomes. Herring (2014) describes a successful collaboration at Ohio State University between a subject librarian and an ESL instructor to strengthen "an established ESL composition course" (p. 128). In a culturally-diverse environment where students' backgrounds vary enormously, some concepts related to the use of scholarly materials, citations and plagiarism are challenging to transmit. A purposeful collaboration between librarians and faculty makes the teaching and reinforcing of such concepts a less difficult task.

There are online tutorials available for teaching purposes. However, because students come from different cultural backgrounds, they have specific needs that cannot be covered by ready-made tutorials. In a study administered at Zayed University in the United Arab Emirates (UAE), Martin, Birks, and Hunt (2010) found that "most tutorials existing worldwide were considered too text heavy, the language too complex, or the examples used [...] were culturally inappropriate" (p. 70).

In her study of 30 female English as a Foreign Language students in the UAE, Johnston, Partridge, and Hughes (2014) encountered multiple challenges that affected her students' IL experience. First, students' English language abilities had an impact on reading information in English, accessing information in their local environment and language, and translating information to make sense of it. Johnston also found that students utilized teacher-directed techniques to read, understand, and organize information.

Faculty–librarians' collaboration in delivering IL is well documented in the research literature. However, with the advent of the flipped classroom modality as an innovative content delivery model, studies that examine its effectiveness in an ESL course that integrates IL skills are rare, especially in a classroom of Arabic speaking students pursuing a medical degree. Hence, this research aimed to fill this gap and provide some insight into the strategies utilized in the design and implementation of the IL curriculum embedded in the ESL course. The study also investigated students' perceptions of the effectiveness of this integration.

INSTITUTIONAL CONTEXT

Supporting the mission of Weill Cornell Medical College in Qatar (WCMC-Q), the Distributed eLibrary's (DeLib) mission is to deliver diverse information resources and services that engage students, faculty and staff of WCMC-Q, as well as local, regional, and international communities. In alignment with its mission, a goal in DeLib's strategic plan is to build a progressive and academically robust information literacy program that supports lifelong learning. DeLib's strategic plan charts the future for information resources and services at Weill Cornell Medical College in Qatar (WCMC-Q) and the Library's role in supporting the biomedical community in the areas of teaching, clinical care, research, and service as well as its wider role in the region and beyond. The DeLib IL plan was designed in accordance with this strategic plan. To provide effective IL instruction to WCMC-Q's users, the DeLib instruction program set two goals. First, it aimed to ensure all students had IL skills to the extent required by their programs/careers, and second to ensure that skills were grounded in real need, practiced over time, and taught in a sequential manner to build knowledge and skills without unnecessary redundancy.

Researching available educational literacy models, we found several that could inspire our needs. Examples of these models include The Big6 (Eisenberg and Berkowitz, 1990), Information Search Process (Kuhlthau, 1989), Stripling and Pitts Research Process Model (1988), Pathways to Knowledge Information Skills Model (Pappas and Tepe, 1997), The SCONUL Seven Pillars of Information Literacy Core Model

For Higher Education (Society of College, National,, & University Libraries, 1999), the Association of College and Research Libraries (ACRL) Information Literacy Competency Standards for Higher Education (American Library Association, 2006), and others.

After careful analysis of these various models, DeLib decided to adopt the Association of College and Research Libraries (ACRL)'s IL Competency Standards for Higher Education as a model for its instruction program for premedical and medical students. We chose this model because it was more closely aligned with our current approach to IL instruction, as well as specifically targeting libraries in higher education. As the ACRL is the largest division of the American Library Association (ALA) and WCMC-Q is an American educational institution, it seemed an appropriate partnership.

The ACRL IL Competency Standards model is comprised of five standards that outline specific performance indicators and outcomes. The five ACRL IL Competency Standards (American Library Association, 2006) define the information literate student as being able to:

- S1: determine the nature and extent of the information needed.
- S2: access needed information effectively and efficiently.
- S3: evaluate information and its sources critically and incorporate selected information into his or her knowledge base and value system.
- S4: use information effectively to accomplish a specific purpose.
- S5: understand many of the economic, legal, and social issues surrounding the use of information, access and use information ethically and legally.

To determine our current position, DeLib conducted a curriculum mapping — an analysis of course content to look for the best places and methods to ensure that all students attain the information knowledge and skills needed to progress throughout their education and into their careers. The library instruction coordinator created a list of courses where librarians had been invited to provide instruction. Topics taught in those sessions were aligned to the ACRL IL Standards. This provided librarians with a visual aid in determining the gaps in their IL program (see Appendix ATable A.1) and helped as a basis to develop a new detailed logic model that would fill these gaps and align the IL sessions' content to the EAP course content (see Appendix A,Table A.2).

MATERIALS AND METHODS

The study aimed to inquire about the effectiveness of delivering IL skills content through an EAP course using online tools. The purpose of this study was to assess the IL needs of newly-graduated high school EAP Arab students enrolled in a first-year premedical program at a U.S. private college; to evaluate the effectiveness of IL modules in improving students' research skills as established by the ACRL's IL standards; and to explore first-year college students' attitudes towards the integration of IL skills in an ESL curriculum using the flipped classroom pedagogical model.

The study utilized seven different data collection tools including a needs analysis, quizzes and assignments, search logs, two exit tests referred to as "English for Academic Purposes-Information Literacy" (EAPIL), a survey and focus groups. Qualitative and quantitative methods provided complementary, valid and reliable results.

A needs analysis was first administered at the beginning of the academic year. Ten structured, closed, multiple-choice questions were uploaded in Canvas to measure the initial IL skills of the study participants and provided the basis of this study. Although some consider that administering structured questions forces participants to choose an answer rather than being genuine and deliver their own interpretation of a subject, it was decided to have fixed responses as it would help reach the main purpose of these sessions and cover the IL skills in question. Every question included an additional field for "I don't know" to prevent participants from trying to guess the right answer and thus alter the results. The questions covered the five standards of IL set by ACRL (see Appendix C).

Then all modules included quizzes and/or assignments to assess the students' understanding of the covered topics. These can range from three simple questions to more elaborate assignments where students had to submit an annotated bibliography, describe a search on one of the databases, upload an evaluation sheet for websites, and submit an interlibrary loan request through ILLiad, the DeLib's resource sharing management software. While quizzes were automatically graded, and students received immediate feedback, the assignments had to be graded by librarians and comments were added when necessary. All quizzes and assignments were administered through Canvas and constituted 10% of the EAP total grade for the first semester and 5% for the second semester.

Search logs were used as a method to document students' search strategies. Students were introduced to search log during IL Module 2/Session 2 and were requested to submit it with every essay they turned in for their EAP course. The search log was designed to help students keep track of the search techniques, relevant keywords and databases utilized, limits and filters applied as well as citations of the most pertinent sources found. Through the use of search logs, the librarians' objective was that students would rethink their search strategies, use all search techniques covered during the sessions and practice creating citations. In order to encourage its use, faculty required students to turn in a search log with every essay.

At the end of the first and second semesters, students had to take the same 50-question exit test made up of multiple-choice and true/false questions. Although the test was not graded for the fall semester as some of the skills were not introduced yet to students, it had three main objectives: first, to identify gaps in the sessions delivered during the fall semester; second, to plan new IL sessions for the spring semester; and third, to compare students' skills after each semester. The five ACRL Standards were reflected with 10 questions per standard.

Focus groups have been used before as a means to evaluate IL instruction. In this study, focus groups were used to give the students a chance to voice their concerns, preferences and approaches regarding IL sessions. Young (1993) observed that the interaction between the researcher and participants is less formal and easier to administer in focus groups where "people tend to be less inhibited than in individual interviews" (p. 391). Focus groups took place at the end of the fall semester. Six questions were used as well as sub-questions to help reiterate and explain the initial ones and stimulate the discussion while keeping it aligned with the main purpose of the study. Participants were divided into three groups facilitated by the three researchers. Time varied between 20 and 30 min for each group. No recording device was used due to cultural sensitivity and participants' preference. Researchers took notes manually.

Finally, a five-minute survey was circulated to all students at the end of the first semester. The survey was used as another data collection tool along with the above-mentioned methods. A mix of closed and open-ended questions was designed to collect structured information. Questions addressed the value of IL sessions according to students, their level of confidence in using search techniques before and after the sessions, and the transferability of the skills learned. Students were also asked to report any confusing or unclear ideas and suggest ways to improve the sessions.

PARTICIPANTS

Participants in this research project were 18 first-year college students enrolled in an EAP course that aimed to improve non-native

English speakers' reading comprehension, develop their academic vocabulary, and enhance their writing of college level essays that require the use of research skills. Students come from a variety of school systems such as the Qatari public system, an American Style high-school curriculum and the British system. 16 Qatari students, one Egyptian, and one Palestinian participated in the study, 12 of whom were females and six were males. Their first language was Arabic, but there were two students who also spoke Urdu and Kurdish. Their ages ranged between 17 and 19. In addition to their EAP course, students were enrolled in physics, chemistry, math, biology, and a global health seminar. English was the medium of instruction in all courses. During orientation, all students received a laptop and attended a training session on Canvas, the LMS adopted by the college. Although 21 students enrolled in the Foundation class, three students dropped the course before the beginning of the fall semester and one student dropped the course in the second semester, which brought down the number of participants to 18 in the fall semester and 17 in the spring semester.

STUDY RESULTS

Research question 1

R1: What were the IL needs of newly-graduated high school ESL Arab students enrolled in a first-year premedical program at a U.S. private college?

A needs analysis was designed and administered as a tool to measure students' skills prior to any session. Data were gathered from 10 questions to generate a logic model (see Appendix A, Table A.2) that helped in planning the content of each module and session. 18 students participated in this analysis, which consisted of 10 questions (seeAppendix C).

The needs analysis yielded interesting results that shaped IL sessions' content and provided a targeted learning experience. Results of the needs analysis generated the following goals:

- Reinforce existing concepts by giving more hands-on activities and time for reflection, e.g. academic integrity, plagiarism, access to valuable information sources and search engines.
- Cover new concepts and provide real examples for better understanding and assimilation of new research skills, e.g. citation, Boolean operators, scholarly sources and evaluation of online resources.

During the fall semester, nine sessions were delivered to students in a class setting. Hands-on activities were designed to reinforce the learning experience of the students, engage them in class activities and encourage discussion. This course integration consisted of:

- Eight online modules created and uploaded into the EAP course in Canvas
- Nine weekly 50-min sessions
- Weekly quizzes and assignments

One week prior to each session, module content was created in Canvas including readings, video tutorials, quizzes and assignments to be completed and submitted before class time. During in-class sessions, students were divided into groups of three and were given activities to prepare and share their findings by the end of the session with classmates and librarians. Quizzes, assignments and in-class activities were given 10% of the EAP total grade. Each module and session covered two to three outcomes such as academic integrity, different types of information sources, advanced search techniques, Discovery tool, plagiarism, citation, annotated bibliography, popular, scholarly and peer-reviewed publications, information evaluation, Google Scholar and interlibrary loan service (see Appendix B, Table B.1).

For the spring semester, five sessions were designed and delivered to students every other week in a class setting with hands-on activities, quizzes, assignments and class discussion. The spring semester course integration consisted of:

- Four online modules created and uploaded into the EAP course in Canvas
- Five 60-min sessions

Students' activities including quizzes, assignments and in-class participation were given 5% of the EAP total grade for the spring semester. Modules and sessions covered the following outcomes: find print materials, primary and secondary sources, APA citation style, using the citation management software RefWorks to create, export, and import citation and generate a bibliography (see Appendix B, Table B.2).

Research question 2

R2: How effective were the IL modules in improving students' research skills as per the ACRL's five standards?

Students' final scores on quizzes and assignments for the fall semester were between 72% and 95% for the IL component. In the spring semester, students scored between 78% and 90%.

To have an idea on the quality of references used in students' search logs and essays, students were expected to write five academic essays using scholarly and peer-reviewed references throughout the semester. In interviews conducted with students at the end of the semester, they reported more confidence in using databases and other search tools to locate sources that supported their ideas.

Students also pointed out noticing improvement in using evidence from outside sources, which enhanced their writing quality and developed their background knowledge. Finally, the faculty member teaching the course affirmed students' perceptions and observed that students' essays demonstrated confident use of information retrieved from outside sources and remarkable progress in acquiring IL skills.

EAPIL tests results were used to assess students' competence as outlined by the ACRL standards.

Standard 1

Determining the information required, identifying key concepts and terms that describe the information need and recognizing that existing information can be combined with original thought, experimentation, and/or analysis to produce new information proved to be challenging to the students with an average score of 78%. However, this increased to 80% in the spring semester. As primary/secondary resources had not been covered in the first semester, it was understood to have had an impact. Only 16% responded correctly, while 44% responded correctly in the second semester (see S1-8 in Appendix D).

Standard 2

Results provided some alarmingly low scores, showing the instructors the students' areas of weaknesses. For example, in question S2-9 (see Appendix D), only 11% answered correctly, which was a concern to the instructors who used this information to include learning and activities. It was gratifying to note an increase to 81% on question 9 in the second semester and an overall score of 84% for standard 2. Students showed more confidence in their search techniques. They performed better searches using Booleans, keywords, truncation and synonyms and were able to select the best database to use for each topic.

Standard 3

The first EAPIL test showed that students did very well with being able to evaluate information sources and deciding what would suit their needs, as noted from the scores, with an average of 92%. However, when they re-did the same test at the end of the spring semester, they did not do as well. This could have been due to the wording of the two questions, S3-5 and S3-10 (see Appendix D) and the ambiguity of the multiple choice answers the students could pick from. Powell and Ginier (2013) noticed the same problem when they administered their post-test. They stated that "a few students indicate*less* confidence in their information-seeking abilities on the post-test than on the pre-test" (p. 298). The authors believed that after receiving IL sessions, students "realize that finding high-quality information" is not as simple as they initially thought while taking the pre-test (Powell & Ginier, 2013, p. 298).

Standard 4

Similarly, questions in the other standards provided valuable information in defining gaps in the literacy program. Standard 4 responses showed similar results, averaging 82%. Traditionally, it has been difficult for librarians to measure standard 4 because they do not collect assignments, homework, and other assessments from students. However, an average increase to 86% was recorded on an EAPIL spring semester test, which shows that with more exposure and practical exercises, students would improve on their results. Indeed, students showed more confidence using scholarly publication and started developing critical thinking skills when evaluating information and determining its relevance and reliability.

Table 1. Results of the EAPIL tests.

Standard 1: *The information literate student determines the nature and extent of the information needed.*

	S1-1	S1-2	S1-3	S1-4	S1-5	S1-6	S1-7	S1-8	S1-9	S1-10	S1 Average	n
Fall	100%	94%	83%	94%	84%	55%	73%	16%	100%	83%	78%	18
Spring	94%	100%	75%	100%	88%	56%	56%	44%	100%	88%	80%	17

Standard 2: *The information literate student accesses needed information effectively and efficiently.*

	S2-1	S2-2	S2-3	S2-4	S2-5	S2-6	S2-7	S2-8	S2-9	S2-10	S2 Average	n
Fall	66%	83%	77%	55%	100%	100%	72%	61%	11%	94%	72%	18
Spring	94%	88%	100%	31%	100%	100%	81%	75%	81%	94%	84%	17

Standard 3: *The information literate student evaluates information and its sources critically and incorporates selected information into his or her knowledge base and value system.*

	S3-1	S3-2	S3-3	S3-4	S3-5	S3-6	S3-7	S3-8	S3-9	S3-10	S3 Average	n
Fall	88%	100%	100%	100%	83%	94%	100%	94%	77%	88%	92%	18
Spring	100%	94%	100%	100%	44%	100%	100%	94%	81%	50%	86%	17

Standard 4: *The information literate student, individually or as a member of a group, uses information effectively to accomplish a specific purpose.*

	S4-1	S4-2	S4-3	S4-4	S4-5	S4-6	S4-7	S4-8	S4-9	S4-10	S4 Average	n
Fall	66%	100%	100%	72%	77%	100%	66%	88%	61%	94%	82%	18
Spring	44%	100%	100%	88%	94%	100%	94%	75%	75%	94%	86%	17

Standard 5: *The information literate student understands many of the economic, legal, and social issues surrounding the use of information and accesses and uses information ethically and legally.*

	S5-1	S5-2	S5-3	S5-4	S5-5	S5-6	S5-7	S5-8	S5-9	S5-10	S5 Average	n
Fall	88%	100%	100%	55%	88%	77%	33%	94%	88%	73%	80%	18
Spring	69%	100%	100%	63%	100%	100%	50%	69%	100%	63%	81%	17

Note. n = number of students who participated in EAPIL tests. One student dropped the course for the spring semester. Cells in blue indicate scores equal or higher to 80%.

Standard 5

This remained generally consistent. Questions 4, 5, 7 and 9 had better scores in the second semester but questions 1, 8 and 10 were unsatisfactory, which was frustrating to the instructors as these questions were based on general college knowledge. Table 1below shows the students' scores on the EAPIL test for the fall and spring semesters.

Research question 3

R3: What were participants' perceptions of the IL content as part of the EAP curriculum?

Two instruments were used to answer this research question. First, the class was divided into three focus groups facilitated by the EAP faculty member, one information services librarian and one information services specialist. During the focus group discussions with the study participants, five major questions were asked. The first question dealt with students' overall perceptions about the IL course. It simply probed what they liked best about it. Conversely, the second question asked them what they liked the least about the course. The following question examined whether participants felt that the skills learned in the IL course could be applied in their everyday life, whereas the fourth question addressed the aspect of integrating IL skills in the EAP course. It examined whether students preferred having the skills taught jointly or separately. Finally, question five investigated students' attitudes towards the use of the flipped classroom modality and how they felt about viewing the material ahead of class on their own.

The results of the focus groups' responses yielded three consistent themes among the study participants. The most prevalent trends that emerged were information evaluation, citations, and search techniques.

Most of the participants expressed their concerns about the class schedule as the sessions were taking place by the end of the day, and students would be tired and easily distracted. Another concern was the impact of IL grades on their EAP final grade. Students felt that 10% was considerably high and asked to either decrease it to 5% or make the sessions optional.

Many participants did not see how they could apply IL skills to their everyday life; however, some of them acknowledged using website evaluation skills for online shopping. Participants also indicated that these skills helped them in their Biology and Science papers in general.

During focus groups, although most of the students stated that they did not like having the sessions compulsory, and many admitted that they did not check Canvas and did not read the material before class, most of them found the content to be useful and beneficial.

The second tool was a survey that included nine questions (see Appendix E). Students were asked about their level of confidence in using search techniques. They were also asked to list three skills they thought they developed during IL sessions and to share their concerns and suggestions to improve the sessions.

In Question 1, "Did the librarian make the purpose of the Information Literacy (IL) sessions clear to you?" 100% of the answers were positive.

Although IL sessions were integrated in the EAP online course, IL modules were easily identifiable and accessible to students. Module 1/Session 1 introduced students to IL skills and the expected outcomes (Fig. 1).

In question 2, students were asked to rate the overall value of IL sessions. While only seven students found IL sessions to be "Excellent," 10 out of 18 students rated the sessions as "Good". Only one student found the sessions to be "Fair." Class schedule (late in the day) during the fall semester might have had an impact on students' perception of IL sessions. Some students were concerned about being obliged to attend this course and the effect it had on their total grades. Other students asked to include more activities to make the sessions interactive as they found the class slow-paced and sometimes boring.

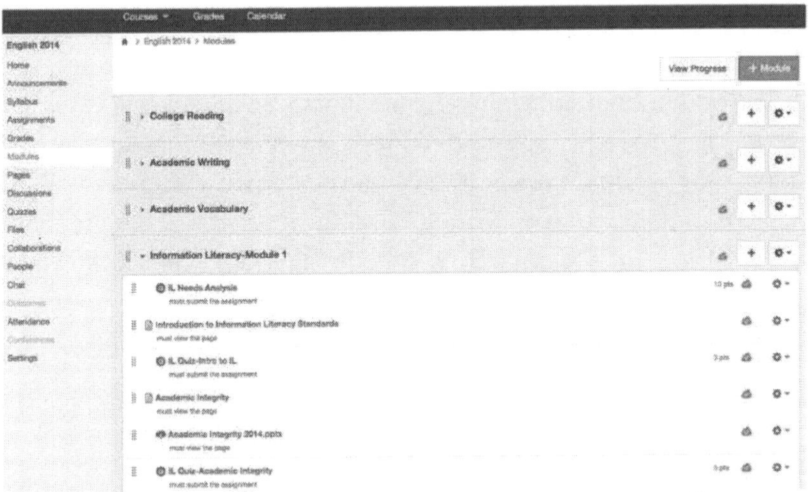

Figure 1. Print screen of Canvas for the EAP online course for fall Semester 2014 showing the name of every module and the content of module 1.

For questions 3 and 4, when students were asked to compare their research skills before and after attending IL sessions, most of them (15 out of 18) felt they were "Not Confident" or "Somewhat Confident" at the beginning of the fall semester. However, by the end of the fall semester, all students felt "Confident" and "Very Confident." Researchers were concerned that students might have chosen to respond positively to question four as they might be trying to please the instructors; however, in our case, the results of the EAPIL test in the spring semester demonstrated that students were indeed more confident using their research skills and became knowledgeable about the breadth of resources available to them. Students also showed some critical thinking skills when evaluating sources.

In question five, students were asked if they would use what they learned from the IL sessions. Students were able to choose multiple answers. This question provided insight into whether students believed they were learning skills that were transferable to real life situations. Fig. 2 shows that students indicated they could use the skills learned in their educational endeavors, especially skills on how to do research for a biology term paper (18), how to find sources for their essays (17), and how to cite when they write a lab report (14). Some respondents also indicated that these sessions would help them in some real life situations such as shopping for clothes and buying a car.

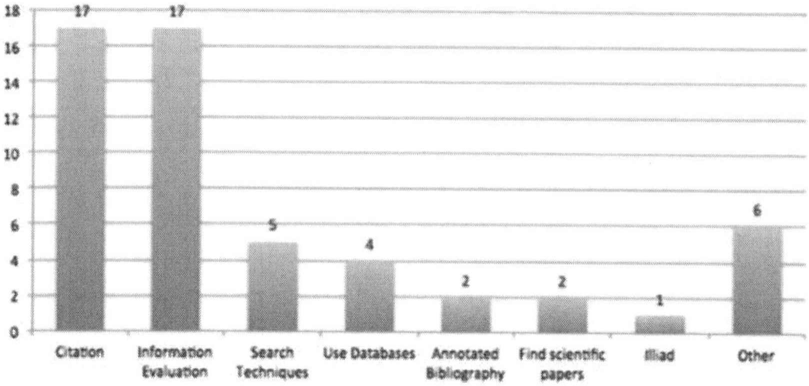

Figure 2. Recurring trends in students' answers to question eight. The answers labeled "Other" were very general such as: "*well prepared search*", "*how to search for scholarly articles*", "*how to collect sources*", "*how to search*", "*how to make a proper search*", and "*how to search for information*".

Question six asked students whether these sessions helped them improve their research skills. Most of the answers (17 out of 18) were positive.

Samples of the answers returned are as follows:

"Yes. I was introduced to new research techniques and how to use databases."

"Yes. Because it helps me a lot in avoiding plagiarism."

"Yes. I did not know which websites to use or how to evaluate them before."

"Yes. I know which sources are credible and which keywords to use."

Positive answers to this question not only indicated to the researchers that student learning was progressing, but it also pointed out which areas students noticed the most growth. Only one student felt that he had not learned anything new because he was already familiar with the skills since he was repeating the course. In fact, this student's pre-test results were high and so were the EAPIL test results.

Question seven, "Was anything confusing or unclear during these sessions? Please explain." While most of the students (14 out of 18) did not find anything to be confusing or unclear, four students expressed some concerns. This could be explained by them reporting not having time to study the modules prior to class.

For question eight, students had to list three skills they learned from the sessions. The researchers decided to divide the answers into recurring trends. Fig. 2 shows the students' answers as per every trend.

Question eight responses yielded clear common topics among students. The most prevalent answer was the use of citations and evaluation of information, which reflects achievement of the goals of instruction as set by ACRL standards. Two other trends that emerged from participants' responses were the use of search techniques and database, which also re-emphasizes a fundamental principle of IL: *finding information.*

Question nine gathered some ideas on how to make IL sessions better. Most of the students preferred to have the sessions earlier in the day (seven students). Six out of 18 students suggested more practice and fun activities. One student suggested having the sessions as optional. Here are some of the answers:

"Time! 4.30!!! It would be more fruitful to have it earlier."

"Have it at an earlier time."

"By making the activities more fun."

"Make it optional and change the time of the class."

EVALUATION OF IL COURSE

Tools for measuring efficiency and effectiveness included:

- Instruction statistics: librarians tracked the number of instruction sessions provided, the number of students, the number of faculty that asked librarians to teach in-class instructional sessions, and any other appropriate criteria.
    - Instructor self-evaluation: librarians regularly evaluated their own teaching effectiveness by examining the results of tests and through regular meetings to determine areas of improvement.
    - Students' survey: at the conclusion of the first semester, students completed a survey on their perceptions of the effectiveness of IL instruction in their EAP course, the results of which were used to improve next semester's curriculum and instruction.
    - With the data received from the evaluation process, DeLib ensures that the IL program addresses ACRL IL Competency Standards for Higher Education, DeLib's Strategic Plan and contributes to WCMC-Q's educational mission as a whole.

CONCLUSION

Before establishing a focused partnership between the ESL faculty and the librarians at our institution, IL instruction was irregular and sporadic. For example, librarians were invited to provide one-shot one-hour sessions that covered multiple topics in a short time without a focus on students' literacy needs. Different classes provided different instructions depending on faculty requirements. Moreover, sessions were mostly lecture-based and did not allot time for practice, assessment, or reflection, and student progress was not tracked from year to year. Instruction was not based on any rigorous standard due to the lack of a conceptual model that shaped a teaching approach based on a careful plan and directional goals. Consequently, this lack of focus resulted in inconsistency and non-uniformity in meeting the IL needs of ESL students.

To find a solution to these learning needs, the ESL faculty and librarians collaborated in the design of an integrated IL plan that adopted the ACRL's five standards. This partnership also helped create a logic model that consisted of inputs, outputs, activities, outcomes and impact of the IL sessions to provide a clear picture of the teaching goals and learning outcomes based on the ACRL IL standards. Faculty support is essential to ensure the success of this collaboration. For example, weekly classroom time was secured as part of the ESL course as well as a percentage of the overall grade was reserved for the IL assessments to maintain student engagement. Furthermore, with the institutional adoption of a new learning management system called Canvas, the IL facilitators were able to utilize the flipped classroom pedagogy through uploading short in-house instructional videos in order to increase students' self-regulated learning and encourage their preparedness.

As a result of these changes and collaborations, students were able to demonstrate transferable skills to the other courses in the program, such as biology, chemistry, global health seminar, and ESL writing. Faculty members teaching these courses expressed satisfaction with students because they were able to use citations correctly, evaluate research resources found on the web effectively, and write using sources more confidently. Students were also capable of using critical thinking strategies to evaluate concepts both in their academic career and their personal lives. Finally, the ESL faculty believed strongly that this was a successful partnership that resulted in students' readiness to tackle more complex material in the medical program thanks to their acquisition of IL skills. In fact, the ESL faculty has incorporated the IL component in his integrated reading/writing curriculum to become a vital component and indispensable part of the foundation program.

This study has demonstrated that a partnership between faculty and librarians was a success. Recently at our institution, basic sciences faculty have approached the librarians to integrate information literacy skills in the biology curriculum for the newly enrolled first year medical students. Due to the limited classroom time available, the flipped modality has become an added avenue to allow time for students to access IL materials at their own pace. To achieve similar results, the authors recommend that institutions of higher education encourage librarians to approach supportive faculty and seek ways to collaborate. A needs analysis should be administered in the beginning of the semester to identify the information literacy needs of students. Having a reliable, user-friendly learning management system (LMS) is fundamental to the success of such a partnership. There also needs to be ongoing communication between the faculty and librarians to monitor students' progress, evaluate the content and improve the delivery of the curriculum. Librarians need to diversify the pedagogical approaches in delivering the content of information literacy to accommodate various learning styles. At the conclusion of the program, faculty and librarians are advised to assess the effectiveness of the program by administering surveys to the students and conducting focus groups to seek students' feedback.

FUTURE DIRECTION

To keep this partnership innovative in the future, the faculty and IL facilitators plan to expand the topics and reexamine the skills covered in their curriculum to meet the needs of students due to impending changes in the medical program curriculum that will take place in fall 2016. To ensure the students' readiness to deal with this curricular change, the IL librarians would need to stay abreast of the changes to meet the needs of students early on in the foundation and premedical programs. In addition, based on the survey and focus groups' results, more integrative learning materials and interactive teaching activities would have to be developed, and more diversified assessments would need to be used on the LMS Canvas. Questions on the EAPIL test used to assess students' acquisition of IL skills would need to be updated and enhanced to reflect future curricular changes. Finally, we will need to make a strong case to make the IL course required and credit bearing. It is important to reiterate the support of faculty and the need for collaborative efforts to ensure a successful partnership with IL professionals in order to educate information literate global citizens.

APPENDICES

Appendix A

Table A.1. Curriculum mapping for foundation and premedical years.

Curriculum mapping foundation & premedical IL sessions 2011–2012

Class	EAP	Writing Seminar 1	Writing Seminar 2	Biology
Year	Foundation	Pre Med 1	Pre Med 1	Pre Med 1
Learning objectives	1. Understand the need to cite references 2. Be able to use your topic to construct a search query 3. Be able to find material in PearlCat 4. Know how to search the Academic Search Premier and Web of Science databases 5. Have some idea of how to use RefWorks	1. Know what an annotated bibliography is 2. Be able to identify sources of high quality information 3. Use a variety of techniques for constructing information queries 4. Use DeLib resources to obtain full the text of articles 5. Generate a bibliography using RefWorks	1. Understand Google vs DeLib 2. Understand risks of research with Wikipedia 3. Become familiar with the DeLib Website; including Databases, Course Reserves & Course Support 4. Be able to place InterLibrary Loans (ILLiad) 5. Appreciate Copyright, cite cite cite!!	1. Understand why we need to cite 2. Be able to provide useful citations of books, chapters, encyclopedia entries, web pages/sites, and journal articles
Assessment/Clicker Questions	In-class exercises: 1. Conduct simple search in Academic Search Premier using topic of their choice 2. Enter references into RefWorks manually 3. Export references from RefWorks	In-class exercises: 1. Conduct simple search in WoS using topic of their choice 2. Enter references into RefWorks manually 3. Export references from RefWorks	In-class exercises: 1. Find "help" etc., on DeLib home page 2. Find a book on "writing" using Pearl Cat 3. Find a book on "English Literature" using the subject search in Pearl Cat 4. Use the Writing Seminar Subject guide to find an appropriate database	1. Students use templates to format some of their references in-class
ACRL Standards covered	Std One Std Two Std Three Std Four	Std One Std Two Std Three Std Four Std Five	Std One Std Two Std Three Std Four Std Five	Std Two Std Four Std Five
DeLib 101 Questions Applicable				Biology Information Lit Quiz and Assignment

Note. This table shows a sample of courses where librarians had been invited to provide instruction and the skills covered according to the ACRL IL Standards. Librarians used this table as a basis to determine the gaps in the DeLib IL program and design a new logic model that would take into consideration students and faculty needs as shown in Table A.2.

Table A.2. Logic model for foundation year.

Logic model foundation year 2014–2015

Inputs	**Activities**	**Outputs**	**Outcomes**	**Impact**
Staff, tools & skills needed to accomplish activities	*Hands-on activities needed to reach the set outcomes*	*Assessment tools that show the evidence of the value of our activities*	*We expect the following changes to occur throughout the academic year (measurable through assessment)*	*We expect the following changes to occur in the next 5+ years (difficult to measure)*
Foundation team including Faculty and Librarians ITS support including Audio-Visual department & Educational Computing Services department. Access to LMS "Canvas" & online materials to build interactive content	Search DeLib discovery & be able to read a call number	Quiz	Be aware of key library services	Students will have developed basic IL skills that cover the need of premedical education & prepare students for their medical career. Students will also develop lifelong learning skills
	Create an account & use the interlibrary loan system & submit an ILL request	Assignment – submit an ILL request		
	Search general & specialized databases & understand the scope of each database	Quiz	Recognize & define an information need. Focus the search scope. Be able to use advanced search techniques & find needed information	
	Use advanced search option in databases & different search techniques: Booleans, truncation, phrasing	Assignment – include search string in search log		
	Focus your topic & brainstorm keywords & synonyms through mind mapping	Assignment – include list of keywords & synonyms in each search log		
	Use the search log to document search strategy	Assignment – submit a search log with every English essay	Document the search strategy & keep track of techniques & results	

Primary & secondary sources of information	Quiz	Recognize different types of information sources
Understand academic integrity & adhere to academic ethical behavior	Quiz	Develop ethical behavior in academic life. Understand academic consequences on plagiarism
Understand plagiarism & its consequences	Quiz	
Recognize the parts of a citation in MLA & APA styles	Quiz	Be able to cite sources properly. Utilize a bibliographic management tool to keep track of resources & documents used. Create annotated bibliographies.
Prepare a citation manually in MLA & APA styles	Assignment – include complete citations of relevant documents in search log	
Export/import citation to RefWorks, manage citations & generate list of references	Assignment – submit a bibliographic list using RefWorks	
Cite sources properly & create annotated bibliography	Assignment – submit an annotated bibliography in MLA style	
Evaluate hoax websites & report/discuss findings	Assignment – compare & evaluate 3 websites	Evaluate information critically
Apply the CRAAP test to evaluate online information	Assignment – compare & evaluate 3 articles	
Use Google Scholar & critically evaluate online content	Quiz	Use free online sources to find reliable information
Compare & determine scholarly vs. non-scholarly	Assignment – find 3 scholarly articles for English Essay	Understand scientific publications

Note. This table shows the changes made to the DeLib IL program. Having a clear picture of the inputs/outputs, activities, outcomes and impact of the IL sessions helped in tailoring IL teaching towards defined goals based on the ACRL IL standards.
This logic model for foundation year will also serve as the basis to plan the logic model for premedical as well medical years and develop IL sessions that would reinforce, complement and expand students' information fluency.

Appendix B

Table B.1. Description of IL modules and sessions for the fall semester.

IL-Module 1/Session 1	Module content	Pts
Students were introduced to IL standards by ACRL and academic integrity	Introduction to IL Standards	
	IL Quiz–Intro to IL	3
	Academic Integrity Readings	
	Academic Integrity PowerPoint presentation	
	IL Quiz–Academic Integrity	3
	What do students think of Academic Integrity?	
IL-Module 2/Session 2		
This module covered different types of information sources. It also described the search strategy and provided useful tools (such as a mind map) to generate ideas and (search log) to keep track of the search.	Types of Information Sources	
	Info Sources Reading	
	IL Quiz–Types of Information Sources	3
	How to plan your search strategy	
	IL–Words to watch for in essay questions	
	Tools to help you plan your search strategy-1–Mind Map	
	Tools to help you plan your search strategy-2–Search Log	
IL-Module 3/Session 3		
This module described the search techniques such as Boolean operators, filters, truncation, and phrasing. Two video tutorials on using DeLib Discovery and ProQuest database were created and posted in Canvas.	Search Techniques	
	IL Quiz–Search Techniques	4
	DeLib Discovery	
	IL Quiz–Discovery	4
	ProQuest–Introduction	
	IL–ProQuest Search	2
IL-Module 4/Session 4		
Module 4 covered plagiarism, citation and bibliography using MLA style	IL-Search Log-Harry Potter	5
	What is Plagiarism?	
	IL Quiz–Plagiarism	3
	Different parts of a citation	
	Why do you need to cite your sources?	
	Example of a book citation in MLA style	
	Example of an article citation from online journal in MLA style	
	When to cite your sources?	
	What is the difference between in-text citation and bibliography?	
	Citation Practice Drag & Drop	
	IL Quiz–Citation & Bibliography	3

IL-Module 5/Session 5		
In Module 5, students were introduced to annotated bibliography in MLA style	Process of creating an Annotated Bibliography	
	Annotated Bibliography	
	Example of an Annotated Bibliography of a Book in MLA Style	
	IL Quiz—Annotated Bibliography	3
	IL Annotated Bibliography	5
IL-Module 6/Session 6		
This module described where to find background information and how to recognize popular, scholarly and peer-reviewed journals	Find Background Information	
	IL Quiz—Research Basics	5
	Popular, Scholarly & Peer-Reviewed	
	Scholarly, Or Not?	
	IL Quiz-Popular, Scholarly & Peer Reviewed	6
IL-Module 7/Session 7		
A video tutorial on using Web of Science and Scopus databases was published in this module. Evaluating sources of information and using CRAAP test* for web sources was another outcome taught in this module.	IL—Web of Science + Scopus	
	IL—Evaluating Sources Checklist	
	IL Quiz—Advanced Search	3
	IL Websites Evaluation	3
IL-Module 8/Session 8		
This last module for the fall semester covered Google Scholar and introduced students to interlibrary loan service	Google Scholar	
	IL Quiz—Google Scholar	3
	What is ILL?	
	How to create an account for ILLiad?	
	IL ILLiad	2
IL-Session 9		
There was no module for session nine as librarians found that evaluating web sources was challenging for students, and half of them did not submit the assignment on time. It was decided to use session 9 to do more in-class web sources evaluation.	Website evaluation: students were divided into groups of three, and each group had to evaluate a website using CRAAP test* and write a summary. A discussion took place at the end of the class where students had then to present their findings.	

Note. Pts = Points for every quiz or assignment.
*CRAAP test is an acronym referring to the general criteria used to evaluate resources: Currency, Relevance, Authority, Accuracy and Purpose. It was developed by the Meriam Library at California State University, Chico.

Table B.2. Description of IL modules and sessions for the spring semester

IL-Module 1/Session 1	Module content	Pts
This module presented Library of Congress classification scheme. Librarians created and uploaded a video tutorial on using Scopus database.	How to find a print book in the Reading Room?	
	How to read a call number?	
	Time to Play!	
	Scopus	
	IL Practice Quiz 1	
	IL Quiz 1	5
	ProQuest vs. Scopus	5
IL-Module 2/Session 2		
Primary and secondary sources of information were covered in this module	Primary vs. secondary sources	
	Primary sources of information	
	Secondary sources of information	
	Research companion video	
	IL Practice Quiz—primary vs. secondary	5
	IL Quiz 2	
IL-Module 3/Session 3		
Module 3 revisited citation and covered APA style	APA (American Psychological Association) style	
	How to cite in APA style?	
	Example of a book citation in APA style	
	Example of a journal article citation in APA style	
	Example of a website citation in APA style	
	Why do you need to cite your sources?	
	When to cite your sources?	
	What is the difference between in-text citation & bibliography?	
	Purdue-APA style formatting	
	IL Practice Quiz—APA citation	
	IL Quiz 3	5
IL-Module 4/Session 4		
In this module, students were given more activities for APA, and a general introduction to citation management tool "RefWorks"	APA practice	
	IL—In-text Citation Quiz	5
	RefWorks	
	ProQuest & RefWorks	
	Create a bibliography using RefWorks	
	Example of a book citation in MLA style	
	IL Quiz 4	5
	IL Bibliography and Search Log-Gender Roles	8
IL-Module 5/Session 5		
This session was divided into 2 parts. In part 1, a Q&A session took place where librarians did a quick review of IL modules for the fall and spring semesters and answered students questions. In part 2, students had to take the second EAPIL test.	Q & A	
	EAPIL Test – spring semester	

Note. Pts = points for every quiz or assignment.

Appendix C

Needs Analysis Results.

Q1: When doing a search, deciding whether or not your original information need has been satisfied requires:

95% of your students correctly answered this question.

Answer 1: a. Refining of the search strategy used	No responses
Answer 2: b. Management of the information found	No responses
Answer 3: c. Evaluation of the information found	18 responses
Answer 4: d. I don't know	No responses

Q2: You are looking up information about a specific type of diabetes in a textbook in the library. The best and fastest way to find information in a textbook is:

100% of your students correctly answered this question.

Answer 1: a. Ask your professor	No responses
Answer 2: b. Searching page by page	No responses
Answer 3: c. Looking in the index of the book	18 responses
Answer 4: d. None of the above	No responses
Answer 5: e. I don't know	No responses

Q3: An abstract is:

89% of your students correctly answered this question.

Answer 1: a. Summary of an article	16 responses
Answer 2: b. Quote from book	One response
Answer 3: c. All of the above	One response
Answer 4: d. None of the above	No responses
Answer 5: e. I don't know	No responses

Q4: Which of the following topics would require current information:

100% of your students correctly answered this question.

Answer 1: a. World War II	No responses
Answer 2: b. Health statistics on breast cancer incidents in Qatar	18 responses
Answer 3: c. History of the Pyramids	No responses
Answer 4: d. I don't know	No responses

Answer 1: a. Obesity AND overweight	7 responses
Answer 2: b. Obesity OR overweight	8 responses
Answer 3: c. Obesity AND overweight NOT overeating	2 responses
Answer 4: d. I don't know	One response

Q6: What section of a journal article typically summarizes the main ideas and findings of the article?

100% of your students correctly answered this question.

Answer 1: a. The text of the article	No responses
Answer 2: b. The person who wrote it	No responses
Answer 3: c. The abstract	18 responses
Answer 4: d. The bibliography	No responses
Answer 5: e. I don't know	No responses

Q7: The following citation,

Liebler, R. (2010). Action and ethics in education. *Academic Ethics*, *8*(2), 153–160 is for:

32% of your students correctly answered this question.

Answer 1: a. Website	No responses
Answer 2: b. Journal article	6 responses
Answer 3: c. Journal	One response
Answer 4: d. Book	10 responses
Answer 5: e. I don't know	One response

Q8: Plagiarism is:

100% of your students correctly answered this question.

Answer 1: a. Not citing your sources of information	18 responses
Answer 2: b. Abstract	No responses
Answer 3: c. Citation style format	No response
Answer 4: d. I don't know	No responses

Q9: When evaluating information resources, you should consider: (please select all answers that apply).
21% of your students correctly answered this question.

Answer 1: a. Author's credentials	14 responses
Answer 2: b. Accuracy of information	15 responses
Answer 3: c. Information currency	8 responses
Answer 4: d. None of the above	No responses
Answer 5: e. I don't know	2 responses

Q10: You can copy and paste image into your own work as long as: (please select all answers that apply).
32% of your students correctly answered this question.

Answer 1: a. You add proper citation	16 responses
Answer 2: b. You found it on the web	No responses
Answer 3: c. You found it in an explicitly copyright-free source	10 responses
Answer 4: d. I don't know	No responses

Appendix D

EAPIL Test Questions.

STANDARD 1 QUESTIONS:*determines the nature and extent of the information needed.*

S1-1 'Discovery', DeLib's catalog, contains all of the following except:

Answer 1: Books.

Answer 2: Scholarly articles.

Answer 3: eJournals.

Answer 4: Laser Disk.

S1-2 If a book that you would like to read is not available in DeLib, what is the best way to obtain the book?

Answer 1: Sign up for an account in RefWorks.

Answer 2: Book a plane ticket to the US to go to the library in New York.

Answer 3: Ask the professor if you can be excused from the assignment.

Answer 4: Put in a request through DeLib's ILLiad service.

S1-3 You have just received a research project from your professor and are unfamiliar with the subject. To begin, where can you quickly find background information on this topic?

Answer 1: An encyclopedia.

Answer 2: A dictionary.

Answer 3: A credible website.

Answer 4: All of the above.

S1-4 You need to research information about X-rays. After some initial searching, you decide to focus your topic on the side effects of X-rays. Therefore, you do the following:

Answer 1: Add publication date as a limit.

Answer 2: Refine your keywords to get more focused results.

Answer 3: Check with your friends to see what they are doing.

Answer 4: Search for synonyms for X-rays.

S1-5 How can you find a specific database in DeLib? (Choose All That Apply).

Answer 1: Type the database name in the DeLib search box.

Answer 2: Click on the list of DeLib Databases and browse the list.

Answer 3: Search ILLiad.

Answer 4: All of the above.

S1-6 In DeLib's 'Discovery', you can search for:

Answer 1: Items in all formats (journals, books, DVDs, etc).

Answer 2: Electronic journals only.

Answer 3: Print and electronic books only.

Answer 4: Student information.

S1-7 When requesting a book by ILL you will be able to attain items from: (Choose All That Apply).

Answer 1: Other university libraries in Education City.

Answer 2: Cornell University Library (Ithaca).

Answer 3: Hamad Medical Center.

Answer 4: Qatar University.

S1-8 Your friend, Abdul, went for a walk yesterday and took a photo of a flower. You decide that you would like to include it in the paper you are writing for your Biology course on herbs and spices. Is the picture a Primary or Secondary source?

Answer 1: Primary.

Answer 2: Secondary.

S1-9 You can access DeLib's resources from anywhere with:

Answer 1: A WCMC-Q username & password.

Answer 2: A Qatar driver's license.

Answer 3: A passport.

Answer 4: A Qatar residence permit.

S1-10 After conducting a literature search, deciding whether or not your information need has been satisfied requires:

Answer 1: Refining the search strategy.

Answer 2: Management of the information found.

Answer 3: Evaluation of the information found.

Answer 4: None of the above.

STANDARD 2 QUESTIONS:*accesses needed information effectively and efficiently.*

S2-1 You conduct an interview with an athlete about the use of steroids in football and wish to use a transcript of the interview in your assignment. This would be considered a secondary source of information:

Answer 1: True.

Answer 2: False.

S2-2 You are working on the following research topic: Diabetes in Qatar, and begin your search with the following search strategy: Diabetes AND Qatar.

However, you find many articles and realize that you only want to concentrate on diabetes in children in Qatar. Which search string will provide the most effective set of results?

Answer 1: Diabetes AND Qatar AND Children.

Answer 2: Diabetes AND Qatar AND (Child* OR kids OR "young adults").

Answer 3: Diabetes and Qatar and kids.

Answer 4: Diabetes Qatar Children.

S2-3 Maria has gathered many citations for her research paper and would like to save them in case she does a similar paper in the future. Which software that is available through DeLib would be her best choice?

Answer 1: Database.

Answer 2: Grab-It.

Answer 3: RefWorks.

Answer 4: Rosetta Stone.

S2-4 A search string used in one database can always be copied straight into another database.

Answer 1: True.

Answer 2: False.

S2-5 Words with identical or very similar meanings are called:

Answer 1: synonyms.

Answer 2: Antonyms.

Answer 3: Interesting.

Answer 4: Allusions.

S2-6 Information found on the Internet is always trustworthy.

Answer 1: True.

Answer 2: False.

S2-7 If you are looking for scholarly, scientific articles, the best database to choose is:

Answer 1: ProQuest.

Answer 2: Scopus.

Answer 3: Humanities International Index.

Answer 4: Rosetta Stone.

S2-8 DeLib's 'Discovery' will not help you find:

Answer 1: The book "The Great Gatsby".

Answer 2: DVD of "House".

Answer 3: The database "Web of Science".

Answer 4: The social media "Facebook".

S2-9 All print books in the Reading Room are arranged by:

Answer 1: Author.

Answer 2: Color.

Answer 3: Call number.

Answer 4: ISBN.

S2-10 You are working on a research paper on the effects of mobile phones on brain health in Qatar. The key concepts from this research assignment are:

Answer 1: Handsets, health.

Answer 2: Mobile phone, brain, Qatar.

Answer 3: Mobile, Qatar, work.

Answer 4: Health, Qatar.

STANDARD 3 QUESTIONS:*evaluates information and its sources critically and incorporates selected information into his or her knowledge base and value system.*

S3-1 Summarizing concepts in your research topic or question can help you find keywords to be used in your search.

Answer 1: True.

Answer 2: False.

S3-2 How would you incorporate more keywords into your search to get more focused results?

Answer 1: Use AND.

Answer 2: Use OR.

S3-3 What section of a journal article summarizes its main ideas and findings?

Answer 1: The list of authors.

Answer 2: The bibliography.

Answer 3: The main text of the article.

Answer 4: The abstract.

S3-4 Which of the following are important when evaluating a journal article?

Answer 1: The year of publication.

Answer 2: Peer reviewed or not.

Answer 3: Author credentials.

Answer 4: All of the above.

S3-5 Ahmed has started a research project on male obesity in Qatar. Which of the following criteria would be most important to him when evaluating journal articles?

Answer 1: Geographic area that the research was compiled from.

Answer 2: Where the researcher works.

Answer 3: How many times the article has been cited.

Answer 4: Length of the article.

S3-6 After conducting a search on narcolepsy, Nina found that there were too many results to work with. Which of the following strategies could reduce the number of results?

Answer 1: Filter search to include only scholarly articles.

Answer 2: Revise the general search to a focused search.

Answer 3: Filter search to include publications from the past 5 years only.

Answer 4: All of the above.

S3-7 You need to research if a particular drug therapy is the best course of action for a condition. What would be the most effective and comprehensive course of action?

Answer 1: Use the information from the first article you find.

Answer 2: Examine articles that both advocate and detract from this course of treatment and evaluate their findings.

Answer 3: Find two articles from "New Treatment Journal" and use the most recent one.

Answer 4: Find two articles from "New Treatment Journal" and use the oldest one.

S3-8 You need to find the most recent information on a drug that has just received governmental approval. What type of source is most likely to give you reliable and up-to-date information?

Answer 1: An electronic book.

Answer 2: A textbook.

Answer 3: A pharmacology database.

Answer 4: A pharmacology subject guide.

S3-9 Which source is most likely to contain bias on research for a new drug that cures multiple forms of cancer?

Answer 1: The New England Journal of Medicine.

Answer 2: The new drugs' company website.

Answer 3: Introduction to Cancer Drug Therapies.

Answer 4: The University of Virginia Medical Journal.

S3-10 Your initial research has introduced you to multiple theories on how to treat anaphylaxis precipitated by insect venom. What might you want to do at this point in your research?

Answer 1: Investigate further by using the new concepts within your search string.

Answer 2: Discard all of your previous research.

Answer 3: Save everything to RefWorks.

Answer 4: None of the above.

STANDARD 4 QUESTIONS:*uses information effectively to accomplish a specific purpose.*

S4-1 Reviewing how you write an essay should:

Answer 1: Occur after you've received a grade.

Answer 2: Occur while you're in the process of writing it.

Answer 3: Occur prior to writing.

Answer 4: None of the above.

S4-2 You have found a 3-page section of a paper that you think perfectly summarizes much of what you want to say in your 5-page essay. You should:

Answer 1: Quote it all and just write 2 pages agreeing with it.

Answer 2: Summarize it in your own words and cite it.

Answer 3: Cut and paste it into your essay.

Answer 4: Summarize it in your own words.

S4-3 Trends in data are usually best communicated with:

Answer 1: Tables.

Answer 2: Footnotes.

Answer 3: Charts.

Answer 4: Cartoons.

S4-4 The most effective presentation model for displaying the findings of a written report to a large group, utilizing MS PowerPoint, is... ?

Answer 1: Use a PowerPoint with only a few points on many slides.

Answer 2: Use a PowerPoint with lots of information on as few slides as possible.

Answer 3: Display a copy of the original written report on the screen.

Answer 4: Use a PowerPoint with only a few points on as few slides as possible.

S4-5 The bullet-point format you have used for a class presentation will be fine for your essay on the same topic.

Answer 1: True.

Answer 2: False.

S4-6 You discover that some of your lab results contradict those reported in a published paper. You should:

Answer 1: Mention the contradiction in your report and try to explain or draw conclusions from it.

Answer 2: Use the results from the published paper as yours must be wrong.

Answer 3: Do not mention the discrepancy in your report.

Answer 4: Change your results to look like the published ones.

S4-7 You want to describe the differences in bacterial counts of water from five different locations. After determining the bacterial count of 1000 water samples from each location, it is best to report:

Answer 1: The average count and statistical variation for each location.

Answer 2: A general description of the results, not mentioning numbers.

Answer 3: All 5000 results according to location with no statistical treatment.

Answer 4: The average count for each location.

S4-8 Many databases have a feature that allows you to save a "search history". Why would you use this?

Answer 1: To avoid repeating searches already done.

Answer 2: To remind yourself of concepts you have already searched for.

Answer 3: To help determine which searches were the most successful.

Answer 4: All of the above.

S4-9 What is the most effective way to describe the steps involved in a chemical reaction?

Answer 1: Detailed verbal description.

Answer 2: Series of diagrams.

Answer 3: Bell curve.

Answer 4: Randomized Control Trial.

S4-10 Sa'ad wants to produce some charts and graphs for a written lab report. To do this he will use:

Answer 1: A ruler and pencil.

Answer 2: Microsoft Office Suite: PowerPoint, Excel and Word.

Answer 3: Adobe Writer.

Answer 4: Adobe Photoshop.

STANDARD 5 QUESTIONS:*understands many of the economic, legal, and social issues surrounding the use of information and accesses and uses information ethically and legally.*

S5-1 Which of the following resources are not paid for by WCMC-Q?

Answer 1: Web of Science.

Answer 2: MLA International Bibliography.

Answer 3: Wikipedia.

Answer 4: ProQuest.

S5-2 If you use a picture from the Internet in your paper, you need to cite where it came from:

Answer 1: True.

Answer 2: False.

S5-3 You are doing a presentation for a class. Which of the following do you need to cite?

Answer 1: A picture.

Answer 2: A poem.

Answer 3: An indirect quote, or paraphrase.

Answer 4: All of the above.

S5-4 Making photocopies of an article for your classmates is acceptable under copyright law.

Answer 1: True.

Answer 2: False.

S5-5 DeLib's resources are only available on campus.

Answer 1: True.

Answer 2: False.

S5-6 Which of the following resources are designed to help you cite correctly?

Answer 1: RefWorks.

Answer 2: ILLiad.

Answer 3: Rosetta Stone.

Answer 4: Discovery.

S5-7 The MLA citation style is more commonly used for:

Answer 1: Humanities.

Answer 2: Social Sciences.

Answer 3: Medicine.

Answer 4: Anthropology.

S5-8 When you are off campus, what can be accessed with your WCMC-Q ID & password?

Answer 1: DeLib's print & electronic resources.

Answer 2: Reading Room printers.

Answer 3: DeLib's electronic resources.

Answer 4: None of the above.

S5-9 You have found an electronic book in DeLib's 'Discovery'. As you are leaving the country right away, you decide to print out the book so you can read it on the plane. It isn't a very big book, only 20 pages. This is legal because you are leaving the country.

Answer 1: True.

Answer 2: False.

S5-10 Which of the following are freely available on the Internet? (Check All That Apply).

Answer 1: ProQuest.

Answer 2: Web of Science.

Answer 3: Supreme Council of Health Website.

Answer 4: Google Scholar.

Appendix E

Survey Results.

1. Did the librarian make the purpose of the Information Literacy (IL) sessions clear to you?

Yes	18 students
No	0

2. Rate the overall value of IL sessions:

Poor	0
Fair	1
Good	10
Excellent	7

3. How do you describe your research skills **before** attending IL sessions?

Not confident	8
Somewhat confident	7
Confident	2
Very confident	1

4. How do you describe your research skills **now**?

Not confident	0
Somewhat confident	0
Confident	8
Very confident	10

5. How will you use what you learned from these sessions? (**Check all that apply**)

Biology research paper	18	Lab report	14
Buying a car	5	Essays	17
Like a post on Facebook	1	Traveling	7
Clothes shopping	12	Painting a picture	1

6. Have these sessions helped you improve your research skills? Please explain

Yes. I was introduced to new research techniques and how to use databases.

Yes. I have been using scientific databases.

No. Because I am already familiar with all the skills since I have taken them before.

Yes. Because it helps me a lot avoiding plagiarism.

Yes. I did not know which websites to use or how to evaluate them before.

Yes. I know which sources are credible and which keywords to use.

Yes. Because it provided information and taught things that I did not know before, ex:annotated bibliography.

Yes. Now, not only do I know how to use databases, I know how to limit my search and by doing so research isn't as time consuming as it was before.

Yes. I learned how to differentiate between resources, how to use them and how to evaluate them.

Yes. 1. Determine the reliable source from what is not. 2. Cite in appropriate way.

Yes. Now I know how to find what I exactly need by using certain search techniques.

Yes. I have learnt to differentiate between scholarly/peer-reviewed articles from less valid articles.

Yes. Because I learned how to research for desired info. From credible websites, while I didn't know how to do that before.

Yes. I now know how to use database and other resources available in Cornell. I know how to cite resources and evaluate websites.

Yes. Because I know how to cite websites and books. Also, I learnt how to evaluate websites and articles to see if they are reliable or not.

Yes. Because now I can choose appropriate resources that are reliable and can be trusted.

Yes. I learned how to cite and how to look for information in any database.

Yes. I learned how to look for the information that I need for my assignments, how to evaluate it, how to use it, and how to cite it.

7. Was anything confusing or unclear during these sessions? Please explain

No

No. If anything was confusing instructors helped me.

Not really, the instructors were helpful in answering and clearing some misunderstandings.

No, because I always ask the librarians when I need anything.

No.

Most things are very well explained in the modules, and things we learn from modules apply on the activities in class.

No. Nothing was confusing. The lecturers were good at explaining the main points.

No :)

No. Everything was well explained.

No. Everything was clearly presented. But the activities might be a bit vague as instructions

weren't written and I didn't understand some of the terminology used.

Some activities we did in class we did not know how to do example: citing resources and the order in which names should be in (author, publisher, article...).

Unsure, some of the sessions I had no idea what was going on and some of them I was pretty sure what was the subject.

Yes, there were some materials that I couldn't understand when I read them on Canvas but when we applied them in class I sort of understood what was happening.

No.

We still don't know about ALL what the library provides.

8. List 3 things you learned from these sessions:

1. Boolean, truncation. 2. How to request through ILLiad. 3. Annotated bibliography.

1. Choose right sources. 2. Proper works cited. 3. Well prepared research.

1. Citing. 2. Database. 3. Reliability of source.

1. How to cite my work. 2. How to make a proper research. 3. How to evaluate the sources.

1. Evaluating websites. 2. How to research for a scientific paper. 3. How to cite sources.

1. How to use WCMC-Q resources. 2. How to evaluate websites. 3. Citations.

1. How to cite. 2. How to evaluate websites & resources. 3. How to write an annotated bibliography.

1. How to search. 2. How to evaluate websites. 3. How to cite!

1. How to search (keywords) in the database. 2. How to evaluate the information. 3. How to cite my resources.

1. Differentiate between a reliable source from not. 2. Cite properly. 3. Narrow the research choices.

1. Search techniques. 2. Website evaluation. 3. MLA citation.

1. Evaluation of journals/articles/websites. 2. How to search for scholarly articles. 3. How to cite.

1. Citation. 2. How to research for information. 3. Evaluating websites.

1. Citing resources. 2. Evaluating websites. 3. Using databases-limiting research.

1. Cite. 2. Evaluate. 3. Use databases.

1. How to collect resources. 2. How to cite. 3. How to evaluate websites.

1. Citation. 2. Evaluation of a website. 3. Looking for an information for a scientific paper.

9. How can we make the Information Literacy class better for future students?

Have shorter group discussions.

Giving students more exercises during class.

Make it optional and change the time of the class.

I think it should be combined with English class.

Having these sessions earlier in the day instead of 3:30.

Focus more on reviewing modules rather than the activities.

Having them at early times & having to read the modules or take quizzes in-class/other things discussed with you today :) Thank you.

Just the timing was inconvenient.

By adding more practical practice during the sessions.

Time! 4.30!!! It would be more fruitful to have it earlier.

More movement and exercises.

By making the class more active.

Less boring, shorter, more movement activities that involve movement. No modules to read at home.

Have it at an earlier time.

Change the timing, because if the session was in the evening the student would be already tired.

By making the activities more fun.

Change the timing & more physical activities.

By not having a lot of burden and unaffected grades. Also ensuring that students are learning.

REFERENCES

1. American Library Assocaition (2006). Standards for libraries in higher education.

2. Arnold-Garza, S. (2014). The flipped classroom: Assessing an innovative teaching model for effective and engaging library instruction. College & Research Libraries News, 75(1), 10–13

3. Bowles-Terry, M., Davis, E., & Holliday,W. (2008). 'Writing information literacy' revisited: Application of theory to practice in the classroom. Information Literacy and Instruction, 49(3), 225–230.

4. Bowles-Terry, M., Hensley, M. K., & Hinchliffe, L. J. (2010). Best practices for online video tutorials: A study of student

preferences and understanding. Communications in Information Literacy, 4(1), 17–28.

5. Bruner, J. S. (2009). The process of education. Boston, MA: Harvard University Press.

6. Calkins, S., & Kelley, M. R. (2007). Evaluating internet and scholarly sources across the disciplines. College Teaching, 55(4), 151–156.

7. Conteh-Morgan, M. E. (2001). Empowering ESL students: A new model for information literacy instruction. Research Strategies, 18(1), 29–38.

8. Dhawan, A., & Chen, C. J. (2014). Library instruction for first-year students. Reference Services Review, 42(3), 414–432.

9. Eisenberg, M., & Berkowitz, R. (1990). Information problem solving: The Big Six skills approach to library & information skills instruction. Norwood, N.J: Ablex.

10. Herring, D. N. (2014). A purposeful collaboration: Using a library course enhancement grant program to enrich ESL instruction. The Reference Librarian, 55(2), 128–143.

11. Jackson, S. A. (2014). Students reflections on multimodal course content delivery. Reference Services Review, 42(2), 467–483.

12. Johnston, N., Partridge, H., & Hughes, H. (2014). Understanding the information literacy experiences of EFL (English as a foreign language) students. Reference Services Review, 42(4), 552–568.

13. Kuhlthau, C. (1989). Information search process: A summary of research and implications for school library media programs. School Library Media Quarterly, 18, 19–25.

14. LaGuardia, C. (2011). Library instruction in the digital age. Journal of Library Administration, 51(3), 301–308.

15. Martin, J., Birks, J., & Hunt, F. (2010). Designing for users: Online information literacy in the Middle East. Libraries and the Academy, 10(1), 57–73.

16. McCue, R. (2014). Does a blended learning, flipped classroom pedagogy help information literacy students in the long term adoption of research skills?

17. Middle State Commission on Higher Education (2009). Characteristics of excellence in higher education: Requirements of affiliation and standards for accreditation.

18. Moselen, C., &Wang, L. (2014). Integrating information literacy into academic curricula: Aprofessional development programme for librarians at the University of Auckland. The Journal of Academic Librarianship, 40, 116–123.

19. Pappas, M., & Tepe, A. (1997). Pathways to knowledge: Follett's information skills model. McHenry, IL: Follett Software.

20. Partnership for 21st Century Skills (2015). Framework for 21st century learning.

21. Powell, C. A., & Ginier, E. C. (2013). Lessons learned: Year-by-year improvement of a required information competency course. Medical Reference Services Quarterly, 32(3), 290–313.

22. Reime, M. H., Harris, A., Aksnes, J., & Mikkelsen, J. (2008). The most successful method in teaching nursing students infection control — E-learning or lecture? Nurse Education Today, 28, 798–806.

23. Society of College, National, & University Libraries (1999). Information skills in higher education: A SCONUL position paper. London: SCONUL.

24. Stripling, B. K., & Pitts, J.M. (1988). Brainstorms and blueprints: Teaching library research as a thinking process. Libraries Unlimited: Englewood, Colo.

25. Young, V. L. (1993). Focus on focus groups. College and Research Libraries News, 7, 391–394 Jul-Aug 1993.

26. Youngkin, C. A. (2014). The flipped classroom: Practices and opportunities for health sciences librarians. Medical Reference Services Quarterly, 33(4), 367–374.

5
CHAPTER

Today and in Perpetuity: A Canadian Consortial Strategy for Owning and Hosting Ebooks

Tony Horava

University of Ottawa, Associate University Librarian (Collections), 65 University Private, Ottawa, Ontario K1N 6N5 Canada

ABSTRACT

The Ontario Council of University Libraries (OCUL) is a provincial consortium of twenty-one publicly funded universities in Ontario, Canada. A consortially-built platform called Scholars Portal is our digital library for archiving ebook content and making it available 24/7 to university students and faculty. An Ebooks Committee has responsibility for coordinating the consortial acquisition of ebooks, within the context of an Information Resources Committee. This paper discusses the consortial strategy and philosophy for ebook licensing in OCUL, which involve a focus on ownership and local loading rights, for dual purposes of preservation and immediate access. Key processes, tools, and accomplishments of this innovative service model are highlighted.

Keywords: Ebooks; Consortia; Local hosting; Licensing; Ontario Council of University Libraries (OCUL); Canada

CONSORTIA AND THE EBOOK LANDSCAPE

The growing importance of consortia for ebook acquisition is very much in evidence. According to a 2011 report, colleges in the U.S. purchase 38% of their ebooks via consortial agreements (Ontario Council of University Libraries, 2011a and Ontario Council of University Libraries, 2011b). This represents a substantial portion of ebook expenditures, and is higher than for public libraries (30%) or for special libraries (24%). Consortia in the digital age have been viewed originally as buying clubs that focused on negotiating lower prices for group purchases (Allen & Hirshon, 1998) and exerting greater influence in the development of scholarly resources (Alberico, 2002). Collective acquisition of digital content leads inevitably to a greater focus on cost-sharing and its management. Stern (2010a) notes that "Payment models for individual libraries already allow for complex funding options, but consortial funding allocations and reporting will be far more complicated and will require central tracking for annual analysis across the group." This has become a fact of life for consortia acquiring digital scholarly content, and numerous models for allocating costs exist, whether based on central funding based on governmental sources, institutional funding, or a hybrid of the two. Many criteria can be employed to serve as proxies for demand and value. What is crucial is that the criteria be perceived as equitable, transparent, consistently applied, and affordable by the members.

Cost–benefit analyses reflect the value of providing much greater access to scholarly material through consortia and saving the time of the user (Scigliano, 2002). Today, however, consortia are "pursuing complex cooperative collection development strategies, and what's more, content production, hosting, and sharing" (Zeoli, 2011). In this context, consortia are playing a much larger market role in the scholarly communications ecosystem with respect to the development, acquisition, and integration of digital scholarship into the life of academic institutions (and other sectors as well). One can agree with Maskell (2008) that "consortia might be considered not as augmenting and strengthening the role of academic libraries in that cycle, but rather becoming an increasingly powerful intermediary between the publisher and the academic library."

While there are numerous and powerful advantages for leveraging consortial action in acquiring digital content, there are challenges that are inherent to the structure and functioning of consortia. Hazen (2011, 199) observes that,

The internal workings of consortia reinforce the grounds for doubt. These bodies are instruments of their members' collective will, but also are beholden to each participant's priorities and claims. Group decisions are susceptible to lowest-common-denominator, weak-link-in-chain, and divide-and-conquer distortions. Consortia, in their current form, may be equivocal instruments of collective resolve.

Human dynamics come into play in collective decision-making, thus testing the collaborative will and the common strategic objectives of the group. It is safe to say that there is a very wide spectrum of dynamics, goals, and resource capacities in library consortia as they evolve in their responses and strategies to the thorny challenges in the scholarly communications ecosystem. There will always be a tension between institutional and consortial interests; this is one of the trade-offs implicit in any collective action. Consortia carry out an extensive and ongoing dialogue with each other, and learn from best practices and experience elsewhere in the community. Considering how rapid has been the growth and influence of consortia on the marketplace, it is difficult to predict how this will evolve in light of the massive transformations we are witnessing in the publishing industry, in the re-invention of research in the digital age, and in the ubiquitous impact of information technologies on the delivery of education. What is clear is that ebooks are a disruptive force, in the best sense of the term, and that all consortia need to rethink their licensing strategies in relation to the value, purpose, viability, and integration of ebooks into their strategic thinking.

A MYRIAD OF CHALLENGES

In Ontario as elsewhere, the ebook landscape has become a complex and challenging focal point for libraries in recent years. The chronic 'messiness' of ebooks is well-known, such as the difficulties in a common definition and format for ebooks; the challenges around pricing; DRM restrictions on use; lack of simultaneous publication with print; a clearer understanding of ebook use; and the need for a better integration of ebooks into the workflows of researchers. Part chameleon, part revolutionary, and part adolescent, the ebook challenges our assumptions on the nature and use of long-form scholarship in today's world. It is therefore a time of much experimentation and questioning, as libraries try to harness the rich potential of the ebook to support learning and teaching in a sustainable manner.

The range of licensing options has expanded enormously—subscription, purchase, demand-driven, and pay-per-view acquisition models are proliferating. There are Big Deal-style complete collections (though not always complete!); subject collections; and bespoke or customized collections. There is major growth in the availability of backlist titles in digital form. There are workflow issues around the delivery of front list ebook titles, particularly in relation to timeliness. Polanka comments that, "Even when an ebook was made available for purchase, some publishers imposed an additional delay before ebooks could be included in leased collections. Lack of uniformity across the publishing industry causes ongoing problems for aggregators and libraries alike." (Polanka, 2011). There is a wide range of preferences for book use: while a growing number of users are comfortable with ebooks and associated reader technology, a sizeable minority still prefers print books for some or all learning purposes (Revelle, Messner, Shrimplin, & Hurst, 2012). There are vendor integration issues around the impact on approval plans, duplication of content, MARC records, invoicing, and comprehensiveness of coverage. There is the lack of standardization of ISBNs for ebooks. Conference papers, online and in-person conversations in the library world are peppered with concerns on how to address these many challenges—and to respond to user expectations for digital content delivery, ease of use, and timeliness of availability. A search in the Academic Complete Database on the licensing of ebooks[1] yielded 1559 hits in total, as of this writing. The annual number of hits has increased on an annual basis: 270 for 2012, 264 for 2011, and 240 for 2010. Professional dialogue on ebook issues is exploding in many directions, such as patron-driven acquisition, which is "poised to become the norm" (ACRL, 2012) in academic libraries. Experimentation around the forms and functioning of e-textbooks is another indicator of the turbulent state of the venerable monograph. The only certainty is that the future forms of the book will be diverse, media-rich, and predicated on business models that exploit new technologies for delivery, interaction, and discoverability.

Looming above this landscape, however, is the sobering financial cloud—all libraries are struggling to maintain strong collections in the context of financial restraint, which in many cases means a flat or declining budget. The loss of purchasing power constrains the options available for libraries vis-a-vis ebooks. As the serials and continuations budget (mostly spent on digital content) is subject to inflationary pressures, the monies available for monographs is declining and under threat in many institutions. The vendors, encouraged by rapid developments in e-reader technology (e-readers, tablets, and smartphones), as well as a dramatic take-up of ebooks by the general public, are sensing many new opportunities in the ebook market place. The journals market is quite saturated. Hence the publishers and aggregators that are rushing to digitize front list and backlist titles, offering multiple licensing options and delivery channels, and overhauling production workflows so that the digital format becomes the default format of publication. This competitive arena is good for libraries, but it doesn't obviate the need for consortia to carefully assess these options in light of strategic goals and significant challenges in working together with partners in the monograph supply chain.

The Ontario Council of University Libraries (OCUL)

In Ontario, Canada's most populous province, there are twenty-one publicly funded universities. These range from the comprehensive research-intensive university (University of Toronto—over 70,000 FTE) to the small, focused undergraduate university (such as Algoma University—about 1000 FTE) and many others in-between. There are over 435,000 university students in the province. The Ontario Council of University Libraries (OCUL) is the consortium that represents these universities' interests and that spearheads consortial activities. Founded in 1967, OCUL developed collaborative initiatives such as a province-wide interlibrary loan agreement to provide books free of charge, and a union catalogue of serials held in Ontario libraries. Since the late 1990s, a Projects Officer and an Information Resources (OCUL-IR) Committee have coordinated the licensing of digital resources; as of this writing there are about 180 licensed products for an annual expenditure of about 20M. The IR Committee includes representatives from each of the twenty-one member institutions. There are separate specialized groups that coordinate the purchase of digital maps, data sets, and geospatial resources. OCUL works as one provincial consortium within a partnership framework that includes other Canadian regional groupings through Consortia Canada and with other academic libraries through CRKN (Canadian Research Knowledge Network).

This has led to a very unique and innovative form of collaboration "with the goal of enhancing research supports and creating rich learning environments for Ontario's diverse and growing student population." (Ontario Council of University Libraries, 2011a) OCUL supports "Canada's knowledge economy by providing the information tools and access essential for high quality education and research in Ontario's universities." (Scholars Portal, 2011a and Scholars Portal, 2011b). The centerpiece of collaboration is a cyber-infrastructure called Scholars Portal, launched in 2002, which preserves and provides access to a broad range of scholarly content that has been licensed by the consortium on a perpetual access basis (i.e. via purchase agreements with publishers). Scholars Portal was originally developed using seed money from the Ontario government in the late 1990s, and is now sustained by the twenty-one member institutions on a cost-share basis.

Local loading rights have been negotiated to enable the archiving of these resources in perpetuity. As of this writing, there are 33,900,615 full-text articles from 14,084 journals loaded on the journals server; 519,655 ebooks on the books platform (this includes 145,200 commercial titles and 374,455 open access titles digitized by the Internet Archive). Scholars Portal has become a powerful discovery portal for a wide range of scholarly research aggregated on a single platform. This also includes born-digital Ontario government documents catalogued by the Ontario Legislative Library. There is a growing number of social science datasets and geospatial resources housed on other Scholars Portal servers. Scholars Portal also provides services to support library and researcher workflows: it "supports the online inter-library loan platform [RACER] for Ontario's universities and provides support for citation management systems, a virtual chat reference service and other tools designed to aid and enhance academic research in Ontario." (Scholars Portal, 2011a and Scholars Portal, 2011b) The technological infrastructure and the twenty four dedicated staff are financed by the members of the consortium, and located at the University of Toronto, which acts as a service provider to OCUL (while being a member as well). There are three Scholars Portal staff that are specifically responsible for managing and archiving the ebook content that is received from publishers.

Negotiations with publishers for ebook content have been predicated on several key principles: 1) The best possible pricing, such that the consortial agreement is better than what any institution could achieve on its own; 2) Alignment with core licensing provisions as outlined in our model ebook license; 3) Perpetual access rights based on purchase agreements; 4) Local hosting on Scholars Portal; and 5) Prioritization of scholarly publishers' collections based on a consensus of our institutional research and teaching needs. These principles will be explored in the course of this paper.

EBOOK PLATFORM

Scholars Portal Books is our locally-built platform for archiving and accessing scholarly monograph texts to support teaching, research, and learning. It is based on ebrary software purchased in 2009, following an RFP selection process. It is a 'light archive'—it is regularly accessed by authorized users to search and browse the collection. The platform provides the ability to search individual collections, or across all public or subscribed collections. Searches can be refined according to various facets—subject, language, author, series, and type. The interface also provides related journal articles on the search term, thus providing an added benefit for students who may be looking for monograph and/or journal literature. Access to collections is based on entitlements associated with acquisition decisions made by each school, and carefully controlled by IP ranges. Shibboleth authentication is also available for federated identity management and appropriate access. Where required, DRM is in place to ensure that usage of the ebooks respects the terms of the license agreements. The platform allows for the highlighting of text, bookmarking, or saving of books, based on setting up personal accounts. PDF downloads are supported, and export to citation management systems such as RefWorks and EndNote is available. A bilingual interface (English and French) is available, in recognition that several of our institutions are either bilingual or include substantial numbers of French-speaking students and professors in their communities.

Once a consortial license has been signed, the publisher works with staff to transfer data and often metadata (although sometimes we need to get MARC records from a different source). Once staff receive the data, they analyze the files and the metadata to determine if there are any issues to address, such as incomplete or corrupt files, and setting up the entitlements for the participating schools. Once the problems are resolved, one of the programmers writes a loader and the books are loaded. The MARC records with the Scholars Portal URL are generated. The team then does a quality assurance check before distributing the MARCs to schools and opening up access. The lack of standardization of ISBNs for ebooks is an issue, i.e. multiple or problematic ISBNs. There is a wiki that provides ebook files and metadata to the institutions (organized by publisher/collection and date). The process of working with publishers to source and load book files has been a learning experience for everyone involved. The ebook record specifications are provided on the wiki; there are general requirements on record creation as well as MARC field-specific requirements. The cataloguers within OCUL libraries has been very active in reviewing MARC records from publishers, discussing issues on cataloguing of ebooks and developing the above specifications.

It is no exaggeration to say that the Scholars Portal Books platform constitutes one of the most innovative and successful collaborative ventures in the consortial environment. This has been recognized by The Charleston Advisor that awarded OCUL an Ebook Innovation Award (Machovec, 2012). OCUL's collaborative approach to providing its members with access to a growing collection of ebooks through the Scholars Portal Books Platform was cited by George Machovec for offering "greater local control, customized functionality, and permanence that can be depended upon." (2012). Scholars Portal as an organization has won an Innovation Achievement Award from the Canadian Association of College and University Libraries (CLA, 2005).

The Ebooks Committee

In 2002–04, OCUL participated in a tri-sector consortial agreement for NetLibrary titles, through the COOL (Consortium of Ontario Libraries) initiative. This involved the colleges and the public libraries in a unique model of cross-sector collaboration for digital collection development. This was the first large-scale consortial acquisition of ebooks in Ontario. Each consortium was responsible for selecting its own titles, and significantly the pool of total titles was made available to all three sectors. It resulted in 9229 titles being purchased. NetLibrary was problematic in terms of the single user, single book model. The interface at that time was quite primitive, and there were challenges in obtaining access to titles and to MARC records. The coordination of this initiative was undertaken by three librarian members of the OCUL-IR committee as well as the Projects Officer. This experience—early in the evolution of the ebook marketplace—revealed the need for a designated group to oversee consortial offers for ebook collections. There was a multiplicity of questions around acquisition and licensing models, delivery channels, pricing, access methods, content availability, and MARC records.

The transition from print books to ebooks has been a long series of growing pains, as publishers, vendors, and libraries have struggled to define a sustainable scholarly supply chain that can meet everyone's interests and workflow capabilities. As is well known, this has occurred in marked contrast to the journals industry, where the transition from print to digital has been relatively painless and well-accepted—the e-format is now the stable, standard model for production and delivery of journal literature. For many reasons—such as immature business models, licensing restrictions, usability problems, and lack of widespread availability of titles—this has not been the case for monographs in digital format. At the same time, there was growing interest in exploring and resolving these issues in order to advance the development of the digital library. As more academic libraries embraced strategies for enhancing their web presence and for creating, acquiring and managing digital scholarship, this became more important.

In recognition of this reality, OCUL-IR struck an Ebooks Committee in 2007. The committee's mandate is to play a leadership role in ebook acquisition and licensing strategies. This creates efficiencies in the relationships with publishers and vendors for acquiring new content. The committee is composed of six members: two from small institutions (under 10,000 FTE); two from medium-size institutions (10–25,000 FTE), and two from large institutions (above 25,000 FTE). The Assistant Director of Scholars Portal (Collections and Digital Preservation) is also a member. There are two co-chairs drawn from this group. This composition ensures that any vendor or publisher ebook proposal is reviewed by members representing a wide range of interests and circumstances. Moreover, there was a recognition that ebook offers were becoming more and more time-consuming to address. The committee serves a key coordination role in shepherding an ebook offer from the initial proposal, through to the call for interest, the fielding of questions and the seeking of clarifications from the publisher, the crafting of the final offer, the request for decisions, and then hopefully finalizing a license agreement with the vendor.

One of the committee members is designated as the lead negotiator, and works in collaboration with the Projects Officer. The committee examines ebook offers and responds to solicitations of interest from vendors. Through analysis and exchange of ideas, the committee determines whether a proposed offer or a solicitation of interest is worthy of being presented to the membership as a consortial proposal. There is regular consultation with members through a distribution list, where questions are raised and clarifications given. This allows for a consistent process, builds expertise, and ensures that Scholars Portal hosting issues are taken into account (as described above). Monitoring national and international developments, such as ebook business models, innovative consortial opportunities, and intelligence gathering in the publishing and scholarly communications field, are also important to this process. The committee formally reports to the OCUL-IR group, at biannual meetings, where accomplishments and strategic directions are discussed.

This framework has been very successful in enabling the Ebooks Committee to assess offers in a systematic manner, and to be guided by objectives that are critical to the consortium's strategic goals. It should also be pointed out that the committee can proactively initiate negotiations with an ebook vendor, if there is sufficient interest percolating from the members (at least three institutions demonstrating such interest). This provides for greater flexibility.

The Model Ebook License and Vendor Template

One of the key objectives in negotiation is to pursue conformity with the OCUL model eBook license[2] (OCUL, 2011b). The ebook license, vetted by legal counsel, is predicated upon the standard database license agreement developed in 2006 to represent our technological, legal, and business interests in negotiating for e-content with vendors. The archiving of licensed content is pivotal to the value of crafting our own model license for ebooks (Gillies & Horava, 2009):

the local load provision is the heart of why OCUL needed its own distinct model license agreement. Moreover, as the future of various publisher–library preservation partnerships, such as LOCKSS and Portico, is far from certain and vendor assurances of perpetual access are viewed with some skepticism by librarians, a local solution still provides the most reliable, responsible, and utilitarian option for digital archiving. (p. 110)

The license includes key provisions that affect ebook agreements, such as ownership and perpetual access, local loading rights, MARC records, file format, delivery issues, and usage rights specific to ebooks such as interlibrary loans. There is a clause on copyright legislation to ensure that no statutory rights under Canadian copyright law are eroded under a license agreement. This model license is monitored in relation to our rapidly changing environment and updated as appropriate. We are in the process of creating a separate local load license that would complement the standard ebook license; this will make the administration of license agreements better manageable, since entitlements to local loading can change over time, as institutions decide to join existing licenses. Moreover, local loading can apply to various content types, not only ebooks.

Another important element has been the crafting of an Ebook Template for Vendors[3](OCUL, 2011b) to be sent to vendors with whom the committee is negotiating an offer. This is a series of questions intended to capture all the relevant information needed to assess the offer. It also signals to the vendor that we have carefully considered our principles and requirements for a consortial ebook agreement, thus pro-actively articulating what we require and value from a vendor. The template is therefore a reflection of our practices and experience in negotiating ebook license agreements. The vendor is asked to complete this template, which is distributed to members with the offer itself. This is especially important for negotiating with a new vendor, i.e. one with whom the consortium does not have any previous history for ebook acquisition. It is generally not necessary for a vendor with whom we are negotiating a successive ebook agreement.

The template requires that the vendor's offer be in accordance with the OCUL ebook model license. This is a critical issue for us, since this license is our common blueprint for shared principles and interests in regards to ebook licensing and management. As mentioned earlier, an ownership model with perpetual access and local hosting rights on Scholars Portal is a key priority for us. The template poses detailed questions around pricing, such as list price, consortial discounts, minimal purchase, multi-year pricing, deep discounting for print, and any ongoing fees. There are various questions around content, such as inclusions, exclusions, front list description, backlist coverage, types of works, and enhancements such as tables of contents and cover images. There are questions around access and accessibility, such as formats, authentication mechanisms, printing and downloading options, and checkout terms (where applicable).

The template has evolved in relation to our experience with vendors and our platform and access requirements. We now provide detailed specifications on MARC records, such as the OCLC standard and the need for a unique identifier mapping the ebook file to the MARC record. In regards to the issue of accessibility (in the context of the Ontarians with Disabilities Act) we ask questions about compliance with these evolving standards. As authentication systems are providing alternatives to proxy server recognition, such as Shibboleth, we ask for the vendor's technical capabilities in this area. As researchers are becoming more interested in large-scale computational analysis of scholarly data, we are asking the vendor whether they support data or text mining of their licensed ebook content. These issues are also reflected in the model ebook license.

The consortium has also developed a 'local load toolkit'—this is a document providing OCUL librarians with an overview of the rationale behind local hosting, and discusses specific criteria, issues, and talking points to address with publishers. This is significant because local hosting as an overarching philosophy for the consortium applies to various types of content other than ebooks (such as ejournals, primary source material, data sets, geospatial content, etc.) and many publishers are not familiar with the business implications and operational logistics involved in supporting local hosting of their licensed content. For some publishers, it takes awhile to understand what we are doing and why it is important to us.

On the question of local content management of ebooks, Polanka (2011) observes that,

Some libraries negotiate with vendors to obtain ebook files and host them on local servers. This provides greater control, but requires technological expertise to develop the interface and load content. A vendor may also send content files directly to libraries for archival purposes while at the same time providing access through its interface. Good intentions aside, the files are of little use to libraries without the servers, interface, and technological expertise to deliver content to users. (p. 5)

Developing the in-house technological expertise and the infrastructure to host the content has been a critical priority for OCUL. This has meant an ongoing review of user services, hardware capacity and software requirements, and staffing in relation to the consortium's strategic priorities. As can be imagined, this represents a significant and sustained investment by the twenty-one members for our shared services and collections, today and into the future.

Ebook Accomplishments

The Committee has enjoyed much success in negotiating ebook agreements with various publishers that support ownership and local hosting arrangements. The first was a comprehensive front list agreement with Springer, and this set the standard for other agreements. These have included: Elsevier, Oxford, Taylor & Francis, SAGE, OVID, Wiley, SPIE, IEEE, Emerald, Morgan & Claypool, Thieme, and Brill. In some cases, this has not been comprehensive coverage of ebooks (based on what has been available). Institutions can decide whether to participate in a given agreement; this is an opt-in model for licensing. There is much content from Canadian publishers, as described further below. The model ebook license has served as a foundation for negotiations. This has been a milestone in articulating and asserting our licensing interests to provide the broadest possible access rights, legal protections, and requirements for perpetual access and local hosting. The model license has been deployed for ebook agreements, sometimes with only minimal changes required by the publishers. This depends on the dynamics of negotiations. Publishers have learned that we have core requirements if they wish to do business with us. As the scale of licensed content on Scholars Portal has grown, and our model has become widely known, publishers have become more familiar with our interests and requirements. The fact that there are robust security protocols, and that there have been no breaches such as systematic downloading of content, has been strong testimony to the viability of our approach.

There has been a national Canadian licensing agreement (via CRKN) for ebook backlist collections from Oxford, Cambridge, and Taylor & Francis that include the option of local hosting. A total of 8141 titles were acquired in 2008. These titles have been loaded onto Scholars Portal, and are accessible by the schools that have participated in this agreement with Ingram. There has been a patron-driven acquisitions project in 2010 with ebrary, involving sixteen of the OCUL institutions. This led to the purchase of 467 scholarly titles for the participating schools, and provided valuable experience in understanding the strengths and limitations of a consortial PDA project (Davis, Jin, Neely, & Rykse, 2012). These titles generally provided very good value for money, based on the tiered contribution model for financing the project. These titles have been loaded on the Scholars Portal platform. Also noteworthy is a pilot project in 2011–12 with Oxford for the UPSO (University Press Scholarship Online) front list collection—these ebooks are automatically delivered to Scholars Portal upon publication, for local hosting. As a consequence, the digital copy becomes the default format for supporting research and teaching. There were thirteen institutions that participated in this project. We look forward to evaluating this new delivery model for monographs, particularly in humanities and social sciences, and moving forward with the second phase of this agreement in 2012–13. University press ebooks, both Canadian and American, are of great interest to us. This is considered to be core research content that is of broad value to our faculty and students. We have struck an agreement (OCUL, 2013a) with the Association of Canadian University Presses and eBound

Canada to acquire a comprehensive collection of over 3000 Canadian scholarly ebooks published from 2007 to 2012, with perpetual access and local loading rights. Scholars Portal will be the exclusive point of access for this collection. This number is expected to grow to almost 4000 titles by 2014. This is a ground-breaking agreement that will allow institutions to access the e-copy as soon as titles are published. Scholars Portal is thereby becoming an integral player in the evolution of the scholarly communications ecosystem in Canada. This will have important ramifications for the dissemination of scholarly Canadian content, and the participating institutions will be assessing the use and integration of these monographs in their communities. As duplication is problematic for libraries, there will be impacts on acquisition practices around approval plans and firm ordering for print titles.

The consortial strategy for managing ebook licensing and acquisition in Ontario has implied an educational process in our relations with vendors. We have invested time and energy in sensitizing vendors to our core principles; we have communicated our priorities and objectives, in the unique context of the Scholars Portal infrastructure and its centrality for how we do business. This has involved a significant amount of information sharing, negotiating and decision-making. Ownership and local hosting of ebooks are cornerstones of our philosophy and a demonstration of stewardship of our licensed scholarly resources. Some publishers have readily accepted this model, while in other cases there was significant discussion needed to convince key players and gain approval. Many discussions on the implications and logistics of hosting ebook files and associated MARC records have occurred. Negotiating these matters has made us more assertive, more experienced, and more focused on our strategic goals.

Moving the goalposts of vendor relations to adopt our frame of thinking has been an important theme during the past several years. The deployment of the model license and ebook template to assert our requirements and interests have been important milestones in this regard. This has mitigated risk for the consortium as a whole and for the individual members—we only agree to terms and conditions for ebook agreements that meet our requirements, and we have walked away from negotiations that haven't lead to a positive outcome. An important spinoff has been the growth of expertise among the committee members in ebook issues in general and the dynamics of negotiation in particular.

What will be the attributes of the scholarly monograph, assuming it survives, in an information culture where so many competing alternatives for disseminating research exist? What will be its intellectual form and structure, what media will it involve, and how will it be used in different and novel ways to address research and learning needs? All players in the ecosystem are seeking a sustainable, forward-looking strategy that can transform the scholarly monograph into a viable form that aligns with the business models, technological options, and supply chain realities of the digital era. We feel that our strategy will safeguard our collective investments in long-form monograph scholarship and serve our patrons effectively into the future. We are proud that in February 2013 Scholars Portal received certification as a trusted digital repository for journals, following an audit process with the Center for Research Libraries, and thereby became the first Canadian library organization to receive this distinction. It is hoped that this certification will eventually extend to ebooks as well.

This milestone demonstrates OCUL's commitment "to the long-term preservation of scholarly resources for the benefit of future students and researchers" (OCUL, 2013b).Stern (2010b) observes that "Preservation of digital and born-digital materials is still a topic of debate. Consortia are the logical groups to explore these details and to develop best practices to ensure safe storage, functionality, and access". Preservation of knowledge and information resources for future generations has been a core value for librarianship, but the ground has shifted dramatically as we leave the print era behind—the quantity and range of scholarly works in digital formats has increased exponentially in recent years, and preservation planning today requires a new paradigm of thinking. We have squarely addressed the challenge of long-term preservation for our community, while building a large-scale digital library of books to be accessed and widely used on a daily basis.

CONCLUSION

Based on a shared vision and technological infrastructure, the OCUL consortium has developed a comprehensive and ambitious strategy regarding ebook acquisition. Ownership, preservation, and integration on our own platform are key goals that have inspired us. Our strategy is one that few consortia can afford to adopt, as it requires a long-term commitment to a shared technological infrastructure, significant staff support, and the ability to share costs in a sustainable manner. We now have a critical mass of ebook content. We have experience in negotiating numerous agreements with publishers, and in the logistics of loading and archiving ebooks and associated metadata. We have learned many lessons along the way and we have developed a collaborative ethos that has served us well. While there are always challenges in working with publishers, and working within a consortial framework, we now have a large measure of control over our collective investment in scholarly ebooks, today and into the future. As the market for monographs rapidly evolves, we expect there will be more opportunities for our Ebooks Committee and for Scholars Portal, whether it be in new 'business' models, new content, new partnerships or developing innovative roles in the scholarly communication supply chain. It is an exciting time to be engaged in the messiness of ebooks.

REFERENCES

1. Alberico, S. (2002). Academic libraries in transition. New Directions for Higher Education, 120, 63–72.

2. Allen, B., & Hirshon, A. (1998). Hanging together to avoid hanging separately: Opportunities for academic libraries and consortia. Information Technology and Libraries, 17(1), 36–44.

3. Association of College and Research Libraries (2012). Top ten trends in academic libraries: A review of the trends and issues affecting academic libraries in higher education. College & Research Libraries News, 73(6), 311–320.

4. Canadian Library Association (2005). Ontario Council of University Libraries (OCUL) awarded 2005 CACUL Innovation Achievement Award.

5. Davis, K., Jin, L., Neely, C., & Rykse, H. (2012). Shared patron-driven acquisition within a consortium: The OCUL PDA pilot. Serials Review, 38(3), 183–187.

6. Gillies, S., & Horava, T. (2009). Developing a model license: A Canadian consortium's experience. In W. Jones (Ed.), E-journals access and management (pp. 91–117). New York: Routledge.

7. Hazen, D. (2011). Lost in the cloud: Research library collections and community in the digital age. Library Resources & Technical Services, 55(4), 199–204.

8. Machovec, G. (2012). From your managing editor: Twelfth annual reader's choice awards. The Charleston Advisor, 14(2), 3–63.

9. Maskell, C. (2008). Consortia: Anti-competitive or in the public good? Library Hi-Tech, 26(2), 177.

10. Ontario Council of University Libraries (2011a). Collaborate, innovate, deliver. http:// www.ocul.on.ca (Retrieved March 2, 2013).

11. Ontario Council of University Libraries (2011b). OCUL model licenses. http://www. ocul.on.ca/node/114 (Retrieved March 2, 2013).

12. Ontario Council of University Libraries (2013a). OCUL, ACUP/APUC and eBOUND partner to promote Canadian ebook scholarship. http://www.ocul.on.ca/node/ 1650 (Retrieved March 2, 2013).

13. Ontario Council of University Libraries (2013b). OCUL's Scholars Portal — Canada's first certified Trustworthy Digital Repository. http://www.ocul.on.ca/node/1639 (Retrieved March 2, 2013).

14. Ontario Council of University Libraries. About. http://www.ocul.on.ca/about (Retrieved March 2, 2013).

15. Polanka, S. (2011). Purchasing Ebooks in libraries: a maze of opportunities and challenges. Library Technology Reports, 47(8), 4–7.

16. Revelle, A., Messner, K., Shrimplin, A., & Hurst, S. (2012). Book lovers, technophiles, pragmatists, and printers: The social and demographic structure of user attitudes toward ebooks. College and Research Libraries, 73(5), 420–429.

17. Scholars Portal (2011a). About. http://spotdocs.scholarsportal.info/display/sp/About (Retrieved March 2, 2013).

18. Scholars Portal (2011b). OCUL overview. http://spotdocs.scholarsportal.info/display/ sp/OCUL+Overview (Retrieved March 2, 2013).

19. Scigliano, M. (2002). Consortium purchases: Case study for a cost–benefit analysis. The Journal of Academic Librarianship, 28(6), 393–399.

20. Stern, D. (2010a). Ebooks: From institutional to consortial considerations. Online, 34(3), 29–35.

21. Stern, D. (2010b). Library use of ebooks 2011 edition. New York: Primary Research Group.

22. Zeoli, M. (2011). How do you eat an elephant? or eContent and the future of the academic book vendor. Against the Grain, 22(6), 59–60.

6

CHAPTER

Libraries Atwitter: Trends in Academic Library Tweeting

Darcy C. Del Bosque, Sam A. Leif, Susie Skar[1]

University of Nevada, Las Vegas, International Academy of Design and Technology - Las Vegas

Keywords: Twitter, reference, Web 2.0, social media, library patrons, academic libraries

INTRODUCTION

Twitter is a social networking platform that allows users to broadcast short messages of no more than 140 characters. This limitation creates the need for quick, to the point messages that must be poignant, as well as informative. Sometimes referred to as microblogging, Twitter is a platform where users can update their followers in short posts distributed by instant messages, mobile phones, e-mail, or the Web (Java et al. 2007). These short messages are called tweets, and often include shortened links for further information.

Although originally launched in 2006 as another tool for friends and family to stay in contact (Aharony 2010), Twitter quickly became a one-stop shop for updates from user's favorite businesses, news outlets, celebrities and more. Despite sometimes being criticized as a useless tool to inform the world what someone had for lunch, Twitter has become a popular resource for businesses and brands to connect with their customers. While some users update their Twitter feeds with information about what they wore today, many use the tool to share relevant information about the world around them and the things they are thinking of at the moment. Many academic libraries have noticed the popularity of the service and have begun to use Twitter to connect with their patrons.

Although often categorized with other social networking sites, Twitter has many unique characteristics. The ability to comment on tweets is similar to the ability to comment on blog posts, however, blogs tend to be more in-depth, since they do not have the 140-character limit of Twitter. Content on blogs also tends to be more longlasting than on Twitter, where posts are rapidly created and then buried as new posts appear. Like Facebook or MySpace, it is possible to use Twitter to connect with individuals; however, with Twitter, instead of "friending," connections are created by "following" accounts. Often, no personal relationship exists between the account holder and the follower, meaning that un-following a Twitter account does not result in any type of conflict. At the same time, there is a need for Twitter feeds to connect with what the follower wants to know, since they will not remain loyal readers solely based on personal relationships.

There are many features of Twitter that make it a technology suitable for use by libraries. Twitter is free, removing any cost barriers for use. Since Twitter relies upon microblogging, it does not require a significant amount of staff time to create and maintain an account. In fact, there are numerous plug-ins that allow pre-scheduling of Tweets, meaning that librarians can establish a weekly or even monthly schedule, further reducing daily staff time needed. Furthermore, many patrons already use Twitter, making it an ideal place for a library to forge connections with them.

Twitter is a fast-paced social network and, depending on the number of accounts a person follows, feeds can move very quickly with old posts being pushed down as new posts come in. Consequently it is easy for libraries to get lost in the conversation. However, there are a number of tools that can be used to facilitate interactions and grab the followers' attention. One such tool is "direct messaging" where a message can be sent to one user and no one else can see the content. This is used much like e-mail and for library Twitter accounts it can ensure the privacy of conversations, while remaining in the informal Twitter milieu. Another feature is called the @reply, or "at reply," which allows users to publicly reply to other users by prefacing the reply with "@username." This keeps posts open for everyone to read, but directs the conversation to specific individuals. In library tweeting, this can be used to welcome new followers or to answer specific questions. A third tool for interactions is called "hashtags," or #'s. These symbols, followed by words unbroken by spaces, allow for the searching of tweets written about similar topics. For example, a library might tweet about banned book week, and end the tweet with the phrase "#bannedbookweek." The phrase becomes a clickable link within Twitter that will search for all the tweets with that hashtag.

This article explores the use of Twitter in academic libraries and identifies underlying trends that exist. While it is understandably exciting to embrace new technology to reach patrons, it is imperative to have an understanding of how to utilize the technology to appropriately connect to library users.

LITERATURE REVIEW

With the advent of Web 2.0, the relationship between the academic library and users has dramatically changed. Tim O'Reilly (2005) coined the term Web 2.0, stating that it involves: "services, not packaged software, with cost-effective scalability; control over unique, hard-to-recreate data sources that get richer as more people use them; trusting users as co-developers; harnessing collective intelligence; leveraging the long tail through customer self-service; software above the level of a single device; [and] lightweight user interfaces, development models, and business models." The capabilities of Web 2.0 enable users to engage the library in two-way communication and knowledge exchanges. Instead of users physically coming to the library, the library delivers services to users via various electronic mediums. Users are also participating in activities that were once the sole purview of the library, such as cataloging via folksonomy, or providing comments on books via blogging. The integration and the utilization of Web 2.0 technology into library services is referred to as Library 2.0. (Casey & Savastinu 2006).

Twitter is among the technologies that libraries are embracing to better connect and communicate with their patrons in the Library 2.0 world. Thomas (2010) noted that libraries, as well as businesses, are already harnessing the momentum of Twitter by creating institutional accounts for marketing, customer service, and other external communications. As Twitter does not have restrictions on institutions or groups having their own user profile, an account is simple and viable for libraries to create and poses many possibilities for use as an outreach and communication tool (Steiner 2009). Although Twitter and other social networks open up new opportunities for academic librarians to reach out to students directly and inform them about the amazing resources at their disposal, Bell (2009) cautions librarians to remember that social networking is not about raising the libraries' profile; it is about students and our commitment to apply technology sensibly to their academic success

All types of libraries are currently using Twitter. For example, public libraries frequently point out highlights on their websites, changes in holiday hours, teen events, and upcoming programs, while university libraries may "tweet" about service issues, upcoming deadlines, library blog posts, and instruction workshops. Twitter gives special libraries a new venue to share information, not just with their internal clients, but also with people outside the institution who are interested in their topics and collections (Milstein 2009, Le Gac 2010, Ho & Hsiang 2010, Bachrach 2010, Dickson & Holley 2010).

With the popularity of Twitter, many authors have proposed ways in which educators, including academic librarians, can engage with their followers. Grosseseck & Holotescu (2008), Milstein (2007), and Mathews (2008) suggested that libraries could share short messages to their followers on Twitter about library events, instructional workshops, book sales, newly available resources, and changes to policies, in addition to providing responses to library user questions. While some users choose to subscribe to these feeds, Twitter updates can also be embedded into the library's homepage to provide instant communication to anyone who visits the site. Additionally, DeVoe (2009) stated that with Twitter, libraries can find out both what their users think about the library and what they do while using it.

Although Twitter is a relatively new social media platform, having recently celebrated its six-year anniversary in March 2012, many academic libraries are already using Twitter in innovative ways to reach their users. For example, the Yale University Science Libraries announce workshops on library resources, provide links to online archives, and give tips on sending text messages to a librarian. Massachusetts Institute of Technology Libraries also provide short announcements regarding workshops, classes, study group information, and other news via Twitter, in addition to interacting with patron posts on the feed (Kim & Abbas 2010).

The New York University NYU Health Sciences Libraries Communications Team created a Twitter account as an outreach tool for students, faculty, and staff to promote resources, events, and relevant news. The team discovered that Twitter can be an excellent way to gather feedback on library services, hours, classes, and so forth. By monitoring and listening to what followers are saying, libraries have the opportunity to fix problems quickly, get user opinions, and essentially listen in on mini-focus groups without having to formally gather users. Although the authors indicated that it is difficult to measure the effectiveness of Twitter, the product is free, and the time and effort to sustain such a venture are minimal. (Cuddy, Graham & Morton-Owens 2010).

Fields (2010) provides an example of how Twitter is used among the reference staff at Koerner Library, the Humanities and Social Sciences (HSS) Library at the University of British Columbia (UBC). Rather than "tweeting" events and services, the purpose of their Twitter account, KoernerRef, is to "tweet" reference questions that will communicate the kind of research being performed by the HSS community at UBC. "Tweeting" the questions develops an archive that can be used for professional development within the department. The questions can be used to discuss how reference interactions were answered, the tools used to find the information, and alternative methods of information access.

Thus far, there have been a handful of research studies on how libraries are using Twitter. In his 2010 study, Stuart collected data from 433 library accounts, including academic, public, state, and national libraries. Although many individual librarians use Twitter, Stuart's focus was on Twitter accounts at the institutional level. Stuart reports that on average 30% of the libraries made one post per day for the duration of their account's life and that 59% of the libraries were following fewer than a hundred Twitter accounts.

Using a sample of 15 public and 15 academic libraries, Aharony (2010) explored the use of Twitter to understand microblogging patterns. The analysis indicated that both types of libraries understand the power of Twitter as a practical channel of communication and attempt to produce a tweet at least once per day. He also noted that public libraries use more informal language in their tweets, compared to academic libraries, perhaps to reach and attract different potential patrons.

Mahmood and Richardson (2011) conducted a survey of the websites of 100 member academic libraries of the Association of Research Libraries and noted that all libraries were found to be using various tools of Web 2.0. Eighty-five libraries in the study were using Twitter for microblogging and were primarily using this technology to share news and announcements.

Cassidy, et al. (2011) surveyed over 6,200 (37%) student library users' utilization and preference of popular Internet and communication technologies at Sam Houston State University (SHSU). Results found that the Twitter microblogging platform has not been heavily adopted at SHSU; barely 21 percent of respondents reported usage and only 10 percent reported an interest in usage. Almost 69 percent of students surveyed had no interest in using Twitter, and nearly 75 percent of students surveyed had no interest in library services using Twitter. Fewer than 20 percent expressed a desire to ask the library questions via Twitter, and slightly over 20 percent wanted to follow Twitter updates on library news, events, and resources. Low respondent interest has reaffirmed the library's decision to use Twitter as a newsfeed tool rather than as a social network.

Mathews (2008) found that academic libraries used Twitter for many practical applications, including assessment, communications, and collaboration with its users. For example, he suggests that librarians can learn by looking at students' "tweets" not only about how they use the library, but also about their general perceptions and expectations. Finally, Matthews notes that Twitter expands the reach of research assistance for librarians. With students using social networks to explain their informational needs, librarians can seek people who need help with resources and services, in addition to syndicating information about events, upcoming workshops, changes in policies, and more. With Web 2.0 applications, such as Twitter, librarians are able to better understand how students use and perceive the library and it adds a new channel for librarians to interact more directly with users.

RESEARCH DESIGN

This study is an intentionally broad exploration of how academic libraries are currently using Twitter. Two hundred and ninety-six academic libraries were randomly selected to study from the 1,460 academic institutions listed in the Carnegie Foundation website[1] . Institutions on this list include Doctoral/Research Universities (Extensive and Intensive), Master's Colleges and Universities (I & II), Baccalaureate Colleges (Liberal Arts & General), and Baccalaureate/Associate's Colleges, located in the 50 states, the District of Columbia, or outlying U.S. territories. A randomized list was generated using Random.Org[2] . The first 296 institutions were chosen from the randomized list to be studied.

Specifically, the study was designed to examine how libraries are currently using Twitter, what their accounts look like, and how Twitter is being promoted on their library websites. Institutional characteristics were gathered from the 2008 Academic Library Survey[3] and the Carnegie Foundation.

Data Collection and Analysis

To gather and compare data consistently, an online form was constructed using Google Forms[4] . The form included 19 questions and an additional space to record any observations about the use of Twitter by the library. To improve accuracy of the results, all three researchers tested the matrix with a sample set of universities before the study began and then compared results. Inconsistencies were noted and minor changes were made to the matrix to make data collection consistent. As a further precaution, during the study two researchers were assigned to each institution and the results were compared at the end of the data collection period. Any inconsistencies were noted and corrected after additional review. This reduced researcher bias, since the information displayed on library websites and on Twitter can be complex and can be interpreted in more than one way.

RESULTS

Nineteen different characteristics of academic library Twitter accounts were reviewed to obtain a broad overview of how libraries are using Twitter, including information about the layout and design of accounts, the number of library account followers, and the number of tweets and type of content they contain. These were compared to additional institutional characteristics of the academic libraries and their parent organizations. Taken as a whole, they provide a glimpse of how academic libraries are currently utilizing Twitter and provide insight into characteristics that make Twitter more successful.

The first question of the matrix asked whether the library had a Twitter account. Other departmental or branch library Twitter accounts were also documented if they were readily findable on the library home page or linked to the main library's Twitter account. Of the 296 schools in the survey, 34% (101 libraries) had Twitter accounts for their main library. Very few libraries had Twitter accounts for individual library departments. In fact, only 10% of schools had other departmental Twitter accounts, for units such as special collections or reference. The researchers expected to find that only libraries who had main branch Twitter accounts would also have Twitter accounts for other departments. Surprisingly, this was not the case, as only 53% of the schools with departmental Twitter accounts also had one for the main library. Of the schools with library departmental Twitter accounts, most often the feeds were for other library branches (47%), reference departments (9%) and archives (9%). Having a branch library did not make a library more likely to have a Twitter account at the main library. Only 8% of libraries with over eleven branches had a Twitter account, while

34% those with 1-10 branches had an account and 34.5% of libraries with no branches had an account.

Regions of the country varied in their adoption of Twitter. In the Rocky Mountains region 66.66% of libraries were using Twitter, with the next highest region being New England at 45.45%.Twitter usage was lowest in the Southwest with only 16.66% of libraries using it, while outlying regions also had low usage with only 22.22% of libraries having Twitter accounts (See Table 1). In addition to regional differences, Twitter usage varied by location. Libraries at academic institutions in rural areas were the most likely to have Twitter accounts (40%), while the least likely to have them were in suburban areas (28.79%). In cities 36.36% of academic libraries had accounts, while 31.58% of academic libraries in towns were using Twitter.

Libraries that had Twitter accounts were more often associated with private academic institutions rather than public institutions. In fact, the only category of institution that had more libraries with Twitter accounts than without were private forprofit organizations. Sixty-two percent of libraries of this type had Twitter accounts, although this was the smallest group of libraries with only eight institutions fitting this category. At private non-profit academic institutions 37% of libraries had accounts, while at public academic institutions, only 27% of libraries had Twitter accounts.

Baccalaureate/Associate's Colleges were the most likely to have library Twitter accounts, with 75% having an account. This group was a small segment of the sample population at 2.73%, however when included with other types of Baccalaureate institutions (Baccalaureate Colleges--Arts & Sciences and Baccalaureate Colleges-- Diverse Fields) this group still was the most likely to have a Twitter account with 39% of the libraries in the sample having one. Master's Colleges and Universities (large, medium, and small) had library Twitter accounts 34% of the time. Research Universities (doctoral/research, very high, and high) were the least likely to have a library Twitter account, with only 23.5% having one.

Libraries with larger numbers of staff did not necessarily have more Twitter accounts than those with low numbers of staff. Those schools with a full time equivalent (FTE) staff between 31 and 40 had the highest percentage of Twitter adoption at 53%, but the next highest percentage of account holders (40%) was among libraries with FTE staffs between just one and ten people. The lowest number of account holders was with libraries that had an FTE staff between 91 and 100, of which none had a Twitter account. There was a tie for the next lowest adoption rate of Twitter (14%) between those with a FTE staff between 61-70 people and those with 71 to 80 people (See Table 2).

Most libraries in the sample offered reference service via e-mail or over the web, only 14 did not (22 libraries did not report on this characteristic). Not surprisingly, a higher percentage of libraries who offered electronic reference had library Twitter accounts (33.6%), versus those who did not offer electronic library reference (21%). No libraries in the study had exclusively electronic collections.

The majority of libraries in this study started their Twitter accounts in 2009, as determined by the first tweet listed in their account (see Table 3). As predicted, the longer a library had an account, the higher the average number of followers that account had. Accounts that were created prior to 2008 had an average of 1,031 followers, while accounts created in 2011 averaged 19 followers (see Table 4). The average number of other Twitter accounts libraries follow is 102, but 68% of the time they followed under 100 other accounts, and surprisingly 12% of libraries did not follow any other account at all.

Two layout and design features that were studied included whether or not libraries used the Twitter logo to link to their accounts and where that logo was placed on the library homepage. Twenty-eight percent of libraries had no logo or link available on their homepage. If they had a logo or link on the home page, it was most often placed at the bottom right (16%), middle right (15%) or bottom middle (11%) of the page. Those libraries that had a logo on the library website did tend to have a higher number of users that followed their account. Overall the average number of followers was 229. Schools that included a logo somewhere on their library home page had an average of 251 followers, while schools that did not have a logo averaged 175 followers.

Customization of Twitter accounts was another design characteristic reviewed. The majority of libraries had at least some identifying information on their Twitter pages. For example, it was found that 90% of the accounts had the library name somewhere on their Twitter page, 78% had the university name, and 52% had photos of the school or library. It was less likely for libraries to have the school or library logo (44%), or to change the color of the page from the default Twitter background (42%).

The number of times a library tweeted in November varied widely (see Table 5). Most libraries (42%) posted between one and ten tweets, with the next largest percentage falling between 11 to 20 tweets (23%). Analyzing the high and low ends of the spectrum, ten percent of the sample accounts did not tweet once in November while 6% tweeted more than 50 times. There was an average of 23 days between the day the Twitter account was analyzed and the last day a tweet was posted; however, this number is skewed by eight libraries that had not tweeted at all this year. If the data is limited to libraries that have tweeted since Jan 1, 2011, the average becomes four days between analysis and most recent tweet, with 27% of libraries tweeting on the day the account was observed; 43% within one day of analysis, 53% within two days of analysis, 61% within three days of analysis, and 65% within four days.

Library Twitter accounts averaged 334 overall tweets from the time the account was created until the day the account was observed. Since the number of tweets varied widely the average or mean are misleading. The mode, or value that occurred most frequently in the data, of 14 tweets is a better indication of how often most libraries were tweeting. It is similar to the average number of tweets per school in November, which was 19. Only six of the 101 libraries had Twitter accounts with more than 1,000 tweets.

Content of Twitter posts was analyzed to determine the primary type of material that was posted. Tweets were further analyzed to determine all secondary content types. The majority of tweeting libraries (55%) used their Twitter feeds to discuss resources. The next most popular primary use was for library events (24%), followed by hours (14%) (see Table 6). The most popular secondary uses for Twitter feeds were hours (24%), library events (19%) and resources and/or campus events (16% each).

The number of followers of library Twitter accounts varied widely. Of the 10 accounts with the most followers, five also ranked in the top 10 for largest number of tweets as well, and eight out of the 10 had an above average number of tweets. There was no direct relationship between the number of followers and how recently the latest tweet had been posted. The 10 accounts that had been in existence the longest had an average of 647 followers, which was more followers than the average account; and the five oldest had an even higher average of 908 followers. Libraries that had at least one tweet in February (the month the study was conducted) had an average number of 261 followers, while libraries that had last tweeted in January had an average of only 58 followers. Libraries that had not tweeted since 2010 had a lower average of 32 followers, while libraries that had not tweeted since 2009 had an average of only 30 followers.

Several features are available in Twitter that allow for interaction between followers and the library. When analyzed, it was found that over half (54%) of the libraries interacted with their followers in some manner. Methods used included, responding directly to a question asked by a follower, by mentioning followers via a directed @ message, or by using a hashtag that clearly indicated their interaction with a follower. Twitter does not allow others to see direct messages between two accounts, so it was not possible for the researchers to measure this interaction. Mentioning, via an @ reply, was one of the least often used feature by followers of libraries. Most of the libraries were not being mentioned by their followers; in fact, only four library Twitter accounts had more than 10 mentions.

Hashtags, a group of words that is preceded by the symbol #, were also used infrequently by libraries. Surprisingly, 54% of the schools never used hashtags, and not a single school always used hashtags. Of the 46% who occasionally used hashtags, most often (91% of the time) the tags were generic in nature, not specific to the school or library. Four percent of libraries used exclusively school-specific hashtags, 18% used school specific hashtags along with other one other hashtag type (generic or libraryspecific), while 5% of schools used hashtags in all three categories.

Another feature of Twitter is the ability to create lists, which allow people to group other Twitter accounts together in a meaningful fashion. Lists can also be used by other tweeters to find multiple accounts that tweet about the same subject that they might want to follow. Ninety-one percent of libraries were on at least one list, while the average number of lists a library Twitter feed was on was 25. Twitter accounts that were listed at least once had an average of 251 followers, while schools that were not included on any lists had an average of only seven followers.

LIMITATIONS AND FUTURE RESEARCH

A primary limitation for this study is the frequency in which Twitter accounts change. Institutions can create or delete accounts at any time. Therefore, it is possible that the characteristics discovered in this study could have changed significantly at any time after the study was conducted. To deal with this limitation, this study focuses on a snapshot of Twitter accounts and library websites during the month of February 2011, noting how many libraries actually had Twitter accounts during that month and quantifying basic characteristics of those accounts, such as number of followers and total number of tweets. To review content of what libraries were tweeting, a different month was chosen for analysis. November 2010 was selected based on the academic calendar, since classes would have been in session at most institutions and it tends to be a busier time of year. These choices helped the researchers to deal with the fact that Twitter is evolving and changing constantly.

Another potential limitation of the study is that the library staff member assigned to manage the Twitter account can also change at any time. This could impact the type and frequency of content posted to the Twitter account. Since only one month of review was completed for content analysis, it is possible that the library may have been more or less active in posting to their account during a different time of year. It is also possible that the choice of month may skew what content was being posted. For instance, there may have been more tweets about hours posted in November due to the Thanksgiving holiday. Future studies could build upon this research by surveying libraries to obtain more comprehensive information about how academic libraries approach using Twitter as a library service or why they have chosen not to create account. Research might focus on the reasoning behind decisions made when posting tweets, including who is chosen to write the tweets or what type of information is tweeted. Other areas for investigation might focus on the decisions behind how and why schools use Twitter, or could focus on students and their opinions on library Twitter accounts.

DISCUSSION

Over a third of academic libraries studied actually had Twitter accounts, but the characteristics of the accounts varied widely. Many libraries were able to accumulate large numbers of followers, suggesting that Twitter could be valuable as a communication tool with patrons. However, other libraries had accounts that were rarely used and offered little value. Much less often, yet troubling, some Twitter accounts seemed antagonistic towards patrons, tweeting things such as, "Why do people insist on coming to the library to watch their online videos? I seriously just saw someone streaming Jersey Shore."

Analysis uncovered interesting facts about the types of institutions that had a higher percentage of Twitter use. For instance, private institutions were more likely to tweet than public, perhaps because of fewer restrictions on Internet use. Additionally, those libraries at academic institutions classified as Baccalaureate/Associate's Colleges were far more likely to have Twitter accounts than Master's or Doctoral level institutions, sparking the question if student focus might be at the cause of this difference. Geographically, several areas seemed to be outliers, either having a far greater or fewer number of Twitter accounts than average, but this did not bleed over to contiguous areas. For instance, the Rocky Mountain region had the highest usage of Twitter at almost 67%, while the Far West had only a 35% adoption.

What to tweet is often a question for new account owners. Over 50% of academic libraries primarily tweeted about library resources. Other main uses of Twitter were for posting hours or for library events. Although some libraries only tweeted one type of content, most had a mixture and often included tweets about campus and community events. The study did not indicate a statistical correlation between content types and number of followers, but it can be noted that libraries tweeting more often had more followers. It may be that libraries posting frequently with content that is useful to patrons attracts and keeps more followers. While on the other hand, not having interesting content or posting too many tweets (and overwhelming a user's Twitter feed) might make it less likely that followers would be engaged in a library Twitter account.

Although many libraries were using Twitter to communicate different messages to patrons, very few were using the resource to carry on a two-way conversation, something for which it is well-suited. Less than 10% of libraries were using Twitter to answer reference questions, although that may in part be due to privacy concerns. Acknowledging comments posted by followers, thanking users for retweeting library messages, and greeting patrons as they sign up to follow the library Twitter account are all ways to make the library Twitter account a welcoming environment.

Some libraries were able to obtain more followers than others. The range of followers was 0-1,040. Schools that tweeted more had more followers, but that was not the only common characteristic for those accounts with high numbers of followers. Many other factors seem to contribute. For example, lists serve a vital function on Twitter, allowing the grouping of accounts that others can use as guide. Many of the libraries in this study were listed with other official institution Twitter accounts, providing a way for students to quickly connect to accounts at their institution. Incredibly, schools that are listed at least once had an average number of followers 3,485% higher than schools that were not listed at all. In general, lists are created by entities outside of the control of libraries, and with the exception of a list created by a parent organization, it may be difficult for libraries to garner the attention needed to be added to Twitter lists.

Libraries with Twitter logos on the library home page also had more followers, with a 43% higher average number of followers than schools that did not have a logo. Another factor related to number of followers was how long the account had been open, with the oldest accounts generally having more followers. Lastly, the number of people the library followed also seemed to make a difference in number of followers of the library. The ratio of number of followers to number of feeds followed was 2.24.

The study revealed that not many libraries are making use of available Twitter tools to enhance their accounts and their interaction with followers. These tools, such as hashtags, direct messages, and @replies, can all be easily used to help make tweeting more effective. For example, despite the popularity of hashtags in the general Twitter public, most academic libraries are not using hashtags. Those that do use hashtags, only use general Twitter hashtags, as opposed to school or library specific ones. By including a library specific hashtag such as #libworkshop, students could quickly and easily find a list of all tweets about workshops in the library.

CONCLUSION

While Twitter has been part of the social media universe since 2006, this study of 296 Twitter accounts from a random sample of academic libraries indicates that libraries have adapted somewhat slowly to this tool. After reviewing the relevant literature on how academic libraries are connecting with Twitter and other social media platforms, it seems that academic libraries are still in the early stages of assessing and evaluating the success of using Twitter and other social networks to connect and communicate with their users.

Only one-third of all academic libraries in this study were using Twitter and only a small number of them were making use of the various tools that would further enhance interactions with their followers. Review of academic libraries using Twitter suggests that that Twitter has the ability to reach and interact with library users effectively by sharing basic information and resources, conversing with patrons, and providing assistance in a quick, yet useful way. As Twitter becomes more widely implemented in academic libraries, future studies will be needed to evaluate how users interact with the libraries via the social networking platform and how libraries' tweets expand over the course of time.

NOTES

1. Available at: http://www.carnegiefoundation.org (accessed 9 February 2012).
2. Available at: http://www.random.org/lists/ (accessed 9 February 2012).
3. Available at: http://nces.ed.gov/surveys/libraries/academic.asp (accessed 9 February 2012).
4. Available at: http://www.google.com/google-d-s/forms/ (accessed 9 February 2012)

REFERENCES

1. Aharony, N. (2010), "Twitter use in libraries: An exploratory analysis", Journal of Web Librarianship, vol. 4, no. 4, pp. 333-350.
2. Bachrach, H.C. (2010), "Being there is only the beginning: Towards more effective Web 2.0 use in academic libraries", Information Science and Technology Directorate Research Report [DTIC-E-TR-2010/06],Defense Technical Information Center, Ft. Belvoir, VA, 2 January, available at: http://www.dtic.mil/cgibin/GetTRDoc?AD=ADA523539&Location=U2&doc=GetTRDoc.pdf (accessed 10 February 2012).
3. Bell, S. (2009), "Is Twitter a must for academic libraries?", Library Journal, 21 May, available at: http://www.libraryjournal.com/article/CA6659755.html (accessed 10 February 2012).
4. Casey, M.E. & Savastinuk, L.C. (2006), "Library 2.0: Service for the next-generation Library", Library Journal, vol. 131, no. 14, pp. 3.

5. Cassidy, E.D., Britsch, J, Griffin, G., Manolovitz, T., Shen, L. & Turney, L. (2011). “Higher education and emerging technologies”, Reference & User Services Quarterly, vol. 50, no. 4, pp. 280-391.

6. Cuddy, C., Graham, J. & Morton-Owens, E.G. (2010), "Implementing Twitter in a health sciences library", Medical Reference Services Quarterly, vol. 29, no. 4, pp. 320-330.

7. DeVoe, K.M. (2009), "Bursts of information: Microblogging", The Reference Librarian, vol. 50, no. 2, pp. 212-214.

8. Dickson, A. & Holley, R.P. (2010). "Social networking in academic libraries: The possibilities and the concerns", working paper [33], Wayne State School of Library and Information Science Faculty Research Publications, Detroit, MI, available at: http://digitalcommons.wayne.edu/slisfrp/33 (accessed 10 February 2012)

9. Fields, E. (2010), "A unique Twitter use for reference services", Library Hi Tech News, vol. 27, no. 6/7, pp. 14-15.

10. Grosseck, G. & Holotescu, C. (2008), "Can we use Twitter for educational activities", paper presented at 4th International Scientific Conference, eLearning and Software for Education, 17-18 April, Bucharest, Romania, available at: http://www.cblt.soton.ac.uk/multimedia/PDFsMM09/Can%20we%20use%20Twitter %20for%20educational%20activities.pdf (accessed 28 January 2012).

11. Ho, H.I. & Hsiang, J. (2010) “Effectiveness of social networking in a university environment”, available at: http://gra103.aca.ntu.edu.tw/gdoc/99/D93922021a.pdf (accessed 10 February 2012).

12. Java, A., Song, X., Finin, T.& Tseng, B. (2007), "Why we Twitter: understanding microblogging usage and communities", paper presented at the 9th WebKDD and 1st SNA-KDD 2007 workshop on Web mining and social network analysis, 12-15 August, San Jose, California, available at:

13. Kim, Y.M. & Abbas, J. (2010), "Adoption of library 2.0 functionalities by academic libraries and users: A knowledge management perspective", The Journal of Academic Librarianship, vol. 36, no. 3, pp. 211-218.

14. Le Gac, M.A.O. (2010). "Twittering libraries: How and why New Zealand public libraries use microblogging", available at: http://researcharchive.vuw.ac.nz/bitstream/handle/10063/1402/thesis.pdf.txt?seque nce=4 (accessed 10 February 2012).

15. Mahmood, K. & Richardson, J.V. Jr (2011), "Adoption of Web 2.0 in US academic libraries: a survey of ARL library websites", Program: electronic library and information systems, vol. 45, no. 4, pp. 365-375.

16. Mathews, B. (2008), "Twitter and the Library: Thoughts on the syndicated lifestyle", Journal of Web Librarianship, vol. 2, no. 4, pp. 589-593.

17. Milstein, S. (2009), "Twitter for libraries (and librarians).",Computers in Libraries, vol. 29, no. 5, pp. 2.

18. O'Reilly, T. (2005). "What is Web 2.0: Design patterns and business models for the next generation of software", available at: http://www.oreillynet.com/pub/a/oreilly/tim/news/2005/09/30/what-is-web- 20.html?page=1 View all notes (accessed 9 February 2012).

19. Steiner, H. (2009), "Reference utility of social networking sites: options and functionality", Library Hi Tech News, vol. 26, no. 5/6, pp. 4-6.

20. Stuart, D. (2010), "What are libraries doing on Twitter?", Online, vol. 34, no. 1, pp. 45- 47.

21. Thomas, L.C. (2010), "Twitter at the office", Journal of Web Librarianship, vol. 4, no. 1, pp. 79-82.

7

CHAPTER

The Renaissance in Library Buildings

Sarah E. Thomas

ABSTRACT

In the enthusiastic embrace of the digital library, with its anytime, anywhere characteristics, some people have assumed that physical libraries will become obsolete. Yet an examination of the behavior of students and faculty and of recent renovation and construction of academic and research libraries shows that brick and mortar buildings are enjoying a renaissance. Libraries have traditionally supported the housing of collections, reader services, and the staff who manage collections and provide services. These three components continue in the 21st century, but librarians, architects, and users are collaborating to produce elegant and functional designs that reflect new spatial allocations and new purposes. Modern libraries incorporate flexibility and comfort to create an environment that is welcoming and that supports a range of research and learning activities. This paper draws on examples from recent construction and reconfiguration of academic libraries in the United

States to illustrate the changing scope of the physical library and the important role it plays in serving as a community hub within a university.

Among the examples selected are several relating to the challenges of collection management, including the use of compact shelving and high-density storage facilities. The paper also traces the evolution of spaces from the catalog card era to the present state in which wireless access is a standard feature. Lastly, the author touches on new and expanded roles for librarians and the ways in which design advances the effectiveness of information professionals in fulfilling these new roles.

INTRODUCTION

As the millennium approached and the promise of digital libraries appeared to be becoming a reality, both the general public and administrators often questioned the future of physical libraries. An article entitled "The Deserted Library" which appeared in the *Chronicle of Higher Education* in November 2001 attracted considerable attention for its descriptions of desolate reading rooms and declining library traffic (Carlson, 2001). University overseers have been prone to imagine that libraries would stop growing and might even shrink, with "everything being digitized." This wistful view of the benefits of technological advancement has been encouraged by the administrators' desire to solve problems of campus congestion caused by relentless construction of new buildings to house expanding collections. This has resulted in many ingenious solutions. At Cornell University the Board of Trustees insisted in 1991 that a new library be built underground to preserve green space and that there be a prohibition on future library expansion on central campus. Libraries have been challenged to use their space more effectively, but they nonetheless are expanding and remaining vital contributors to intellectual discovery and learning.

Scott Bennett, in a monograph published in 2003, notes that in the decade between 1992 and 2001 U.S. library construction projects costing almost $4.5 billion resulted in an annual renovation or construction of library space averaging about 2.8 million square feet. About 40% or 1.1 million square feet was newly built library space (Bennett, 2003, p.9).

Despite the trend toward electronic resources, physical collections continue to grow, especially in the humanities and social sciences. In the sciences, where a higher percentage of the literature appears in digital form, paper backfiles exist increasingly primarily for archival purposes and for professional certification, rather than for active consultation. Libraries have coped with the explosion of knowledge by utilizing compact shelving for materials required for immediate access. More frequently, however, they have opted for the so-called Harvard-style depository, named after the Harvard Depository Library, a high-density warehouse in which books and other materials are stored by size. Retrieval by barcode in buildings reaching to 30 feet high is effected by the use of special forklift trucks. Initial resistance by scholars to the relocation of books and journals to a non-browsable offsite storage facility has been mitigated by efficient delivery mechanisms that bring physical objects to patrons within 24 hours or less or deliver scanned articles to the desktop (Kohl, 2003). A new development, that of regional repositories with shared and integrated holdings from multiple institutions, is taking hold in the U.S. (Smith, 2004). More and more libraries are questioning the practice of storing redundant versions of journals and are moving toward shared and integrated collections to conserve space and increase economies of scale (Reilly and DesRosiers, 2003).

Previous generations of scholars have often conceived the primary purpose of libraries as being the storehouse for books. Indeed, funding for collections and their management remains the largest portion of a library's budget. The construction of book stacks often dominated fortress-like buildings. Access to the collections was channeled through the card catalog, which occupied a place of prominence on the main floor of the library. Adjacent to the catalog cases were the offices for acquisitions and cataloging, since these staff needed to be able to consult the files and bibliographic indices.

INFORMATION COMMONS

Today the online catalog has displaced the card catalog, and libraries have transformed the liberated areas into information commons. With library staff able to access bibliographic information from their workstations, their workspace can be independent from collections and public areas. Institutions are regaining the use of technical services operations for patrons by shifting some staff to office buildings. Harvard, for example, moved technical services to an off-site facility about a mile away from its crowded Widener Library in 2000, renovating workrooms into an elegant and comfortable periodicals and newspaper reading room. Several other organizations have made or are planning similar moves, among them Cornell, Ohio State, The University of Pittsburgh, Stanford, and the University of Virginia. The relocation of staff who do not interact with the public to offices or other buildings frees valuable first-floor or central space for new public and collaborative uses.

Much of new library design or renovation focuses on the services libraries provide. Among one of the most popular installations is an innovation adopted from commercial booksellers such as Borders or Barnes and Noble: the café. Responding to the way society today often blurs the lines between work and home, progressive libraries have introduced cafés with soft seating, specialty drinks like mocha lattes and fruit smoothies, and wireless access. These inviting spaces are magnets for students and faculty alike and serve as meeting places and an informal study environment. They demonstrate the need for human connection, even in a cyberworld. One school, the Rochester Institute of Technology, has designed a series of spaces to foster creativity and community. The Idea Factory, as they have named it, includes a salt water aquarium, a microphone that speakers can reserve to address groups on topics ranging from poetry to public safety, a reading lounge with jazz piped into it, and flexible furniture that can be moved amidst gallery cases displaying student art.

Increasingly, libraries are investing in high quality finishes, solid wood furniture and paneling, and elegant task lamps that provide both an island of warm light and the functionality of electrical outlets and data connections. Readers seem drawn to reading rooms that impart an air of gravitas. Older libraries have refurbished their grand reading rooms as inspirational spaces, while new construction continues to feature modern interpretations of the large halls where students work quietly side by side. These spaces stand in stark contrast to the commotion typical in residence halls or shared apartments, and they impose a constructive discipline of order and serenity.

Students work on a 24-hour clock, and many of them seek study space where they can concentrate and work without interruptions at any hour of their convenience. Consequently libraries are incorporating secure 24-hour study spaces into their buildings. These areas frequently are capable of being closed off from the rest of the library so they cost less to operate during non-peak hours, and they often require an identification card for entrance. The primary purpose of the facility is for study and computer use; there is not high demand for collection access or reference service.

A continuum of spaces from individual to group

Typically, libraries require an array of rooms created to support different work styles and levels of work. Increasingly professors are assigning students collaborative projects. To support this activity, libraries are creating group studies with Internet connections, white boards, and soundproof walls. This category of room is growing in new buildings, and demand remains high. Advanced students conducting dissertation research bring a different set of needs to the library. They seek a quiet desk where they can store books and lock up valuable or personal items. Power and data are standard. Enclosed, locked studies are highly desirable. At the same time that doctoral students advocate for such spaces, however, librarians note a pattern of behavior that makes them hesitate to allocate more single study space. The graduate students work in spurts and spend only a limited portion of their time at their carrels. The dilemma of the library is how to allocate scarce resources. Some institutions, such as Washington University in St. Louis, Missouri, are resolving this issue by creating graduate study rooms with adjacent lockers. A continuum of spaces from individual to group helps target the appropriate space to the user need and moves beyond the rigid, one-size-fits-all character of some earlier library designs.

Faculty studies are also undergoing change. A declining number of faculties make regular visits to the physical library. At Cornell University a recent survey revealed that only 15% of faculty are in the library daily. With ubiquitous access to many resources electronically and competing demands for their time, they are as likely to use library services and collections remotely from their office, from home, or from another location. Studies are often empty, dark and used chiefly as bookshelves. As a result, libraries are rethinking space for faculty, fashioning a flexible set of rooms that can be assigned for different durations, depending on the nature of the professor's work. Here, too, lockers provide privacy and security for working materials, but the rooms in which the faculty work may be assigned on a daily basis or be communal spaces in which multiple desks are available for use, but not permanently assigned to an individual.

In addition to rethinking public spaces to make them more congenial and functional for the modern university student or researcher, librarians are dramatically altering their buildings to reflect new expectations for service or new responsibilities. The insertion of technology into the academic enterprise has changed the way in which people work, bringing activities into proximity that previously rarely intersected. Individuals don't divide their time among reading, email, writing, problem solving, and other activities into separate categories conducted in separate places. Rather, they integrate their efforts and move fluidly from one to another. If, when working at a computer in a library, they have a question, it may relate to software as much as it does to the content or use of an information resource. Some institutions, therefore, have aligned their information technology experts with their librarians, placing them side-by-side. At Dartmouth College the architectural solution to fusing an addition to the main library was to create a street of services. As college residents passed through the building, using it as a thoroughfare to move from one part of the wintry campus to another, they traversed an array of services, including help offered by technologists, reference, circulation, and other desks. The familiar experience of the main street or the shopping mall was translated to the academic setting.

At the University of Chicago three different units in the university, the division of biological sciences combined forces with the science library and the information technologies group to create a computer-enriched teaching and learning center. The Crerar Library has a space that supports individual and group computer activities, multimedia, a visualization classroom, and a cybercafé (Schuler, 2002).

A SUITE OF SERVICES

Other collaborative spaces are becoming prevalent in libraries, particularly in the area of digitization and scholarly communication. North Carolina State University has developed a suite of services that relate to digital scholarship and digital libraries. One floor of their building has been repurposed into the Learning and Resource Center for the Digital Age (LRCDA). Among the operations housed there are a digital media lab, a scholarly communication center, a usability research library, a collaboratory, and several seminar rooms that function as information technology teaching centers and multimedia classrooms.

Another trend in library facilities reflects increasing interdisciplinary work as well as the shift from physical collections to electronic. Many American universities are building combined science or science and engineering libraries that are themselves integrated into science classrooms, laboratories, and workspaces. The University of Virginia has just opened a science and engineering library, and MIT, Columbia, Duke and Princeton all have plans to construct one in coming years. Princeton's facility, a $60 million design by Frank Gehry, will consolidate the geosciences, chemistry, ecology and evolutionary biology, and molecular biology branch libraries and will connect below grade to the math and physics library. It is envisioned to be highly computerized, with very few books, and embedded within the daily life of the scientists and students researching, teaching, and research (Stevens, 2002). It will be collocated with a Digital Map and Geospatial Information Center and the Office of Information technology's Academic Technology Center and the Princeton Institute for Computational Science and Engineering.

Consolidation of collections and services

Just as teams of scientists come together to explore the intersection of their disciplines and collocate in buildings, so do the disciplinary libraries. This consolidation has the added benefit of creating a more effective and efficient delivery of library services since it eliminates redundancy and allows staff to be deployed at fewer service points. The reduction of service points is another theme in library buildings. Users seek one-stop shopping to increase their productivity. In a university where libraries are distributed in several locations, a professor who needs to consult materials in a number of different libraries will spend the equivalent of $70 in retrieving an item from each collection, and the disincentive to interrupt other activities may act as a barrier to seeking out information at all. Consolidation of collections and services saves users of multiple resources time. Librarians are also able to derive advantages from having only one person staff a desk in off-peak hours, rather than three, as in an example at MIT where reference, reserve, and circulation all occur at a single counter.

One of the most advanced plans for a modern library is the Stanford Medical Library and Knowledge Management Center, for which plans are currently on the drawing boards (Pizzo, 2003). Here the Medical School takes full advantage of the freedom and flexibility that technology offers to rethink its traditional library. A significant portion of its print collection will move to offsite storage. Its new services will be split into two facilities, the library and the knowledge management center. The planners envision a distributed learning commons with a casual collaborative core, a visualization computation zone, interactive teaching areas, and live consultations. Librarians will be actively engaged in the full range of the medical educational process, being active in providing information assistance at hospitals and clinics, in classrooms, and in laboratories. The library and its services are dispersed among all facets of the medical school.

Libraries as a font of creativity and inspiration

As these examples illustrate, the face and form of the library today is undergoing significant change. Libraries are shifting from being repositories for information and functioning in self-contained capsules to becoming integrated within the complex interdisciplinary fabric of the university. The overwhelming explosion of knowledge and the challenge of collecting it all physically have served as an impetus for networked library services which function on a worldwide scale. This transition, aided by rapidly evolving information technologies, had enabled librarians to think differently about housing voluminous collections. Libraries today support the flow of information and the exchange of ideas through their services and spaces. Despite the ability of the patron to meet many of her information needs at any hour of the day and in almost any location, library buildings offer unique features that continue to attract a clientele. Libraries are enjoying a renaissance as a font of creativity and inspiration. Through their placement at the very center of the university, they serve as intellectual and social crossroads. Modern libraries are beacons that attract the academic community and act as centers for facilitated knowledge discovery. The buildings, like the organizations that occupy them, are designed to be flexible and collaborative. Far from being deserted and abandoned, they teem with life and excitement. The next generation of students is building its own set of traditions and discoveries that will ensure that the library as place remains a vital part of the learning experience.

REFERENCES

1. Bennett, Scott: *Libraries Designed for Learning.* Washington, D.C. : Council on Library and Information Resources, 2003. CLIR Report 122, November 2003, 89 pp.http://www.clir.org/pubs/abstract/pub122abst.html

2. Carlson, Scott: "The Deserted Library: As Students Work Online, Reading Rooms Empty Out, Leading Some Campuses to Add Starbucks". *Chronicle of Higher Education,* (November 16, 2001). http://chronicle.com/free/v48/i12/12a 03501. htm

3. Kohl, David F.: "Paper and Digital Repositories in the United States". LIBER Quarterly, 13(2003)3/4, p. 244-246.

4. Pizzo, Philip: "Stanford Medicine Information & Learning Environment". The Library as Place: Symposium on Building and Revitalizing Health Sciences Libraries in the Digital Age, November 5-6, 2003, Bethesda, Maryland.http://www.nlm.nih.gov/building/Pizzo.ppt

5. Reilly, Bernard F. Jr. and Barbara DesRosiers: *Developing Print Repositories: Models for Shared Preservation and Access.* Washington, D.C. : Council on Library and Information Resources, 2003. CLIR Report 117, June 2003, 63 pp.http://www.clir.org/pubs/abstract/pub117abst.html

6. Schuler, Peter: "Crerar Library Will Soon House High-tech Computer Laboratory".*The University of Chicago Chronicle,* 19(March 2, 2000)11.http://chronicle.uchicago.edu/000302/crerar.shtml)

7. Smith, Abby: "When Bigger is Not Better". *EDUCAUSE Review,* 39(2004)1, 52-53.http://www.educause.edu/pub/er/erm04/erm0416.asp

8. Stevens, Ruth: "A study in contrasts: Initial plans revealed for Whitman College, science library". *Princeton Weekly Bulletin,* 92(October 21, 2002)7.http://www.princeton.edu/pr/pwb/02/1021/1b.shtml

9. Web sites referred to in the text

10. Learning and Resource Center for the Digital Age.http://www.lib.ncsu.edu/administration/lrcda/index.html

8

CHAPTER

Acquiring E-Books for academic libraries

Hazel Woodward

Hazel Woodward, Cranfield University, Cranfield, Bedfordshire MK43 0AL, United Kingdom

ABSTRACT

This paper outlines the recent work of the UK's Joint Information Systems Committee (JISC) in the area of e-books. The JISC Collections Team is responsible for negotiating deals with publishers and aggregators of e-content for all UK higher education libraries - some 180 in total. In other words it acts as a national consortium for the UK academic community, although it should be noted that all deals are negotiated on an 'opt-in' basis. The JISC Collections Team is advised by a series of format-based working groups - comprising senior academic librarians and library practitioners - including the E-Books Working Group, the Journals Working Group and the Moving Images Working Group. Recently the working groups have formulated vision statements to help inform both their own activities and the education community as a whole. The vision for e-books in UK education is as follows: "The UK education community will have access to quality e-book content that is of high relevance to teaching, learning and research across the broadest range of subject areas. Flexible business and licensing models will support a diversity of needs, allowing users to do what they want when they want and how they want for education purposes. All e-books will be easily discoverable and consistent standards will allow all content to be fully integrated into library, learning and research environments."

INTRODUCTION

This paper outlines the recent work of the UK's Joint Information Systems Committee (JISC) in the area of e-books. The JISC Collections Team is responsible for negotiating deals with publishers and aggregators of e-content for all UK higher education libraries - some 180 in total. In other words it acts as a national consortium for the UK academic community, although it should be noted that all deals are negotiated on an 'opt-in' basis. The JISC Collections Team is advised by a series of format-based working groups - comprising senior academic librarians and library practitioners - including the E-Books Working Group, the Journals Working Group and the Moving Images Working Group. Recently the working groups have formulated vision statements to help inform both their own activities and the education community as a whole. The vision for e-books in UK education is as follows:

"The UK education community will have access to quality e-book content that is of high relevance to teaching, learning and research across the broadest range of subject areas. Flexible business and licensing models will support a diversity of needs, allowing users to do what they want when they want and how they want for education purposes. All e-books will be easily discoverable and consistent standards will allow all content to be fully integrated into library, learning and research environments."

THE JISC E-BOOK STUDY

In 2006 the E-Books Working Group (EBWG) commissioned a feasibility study on the acquisition of e-books by higher education (HE) libraries and the role of JISC (CHEMS, 2006). The study team produced their report in mid 2007 and outlined the following conclusions:

- There is an ignorance in the HE sector of what e-books are available.
- There is low awareness within HE institutions of the value and relevance of e-books.
- Librarians and publishers have a poor understanding of each others needs in the area of e-books.
- Access routes to aggregators' or publishers' platforms are too complex.
- Too few e-books are available.
- Available e-books are not up-to-date or relevant to UK users.
- Pricing models are not appropriate.
- Publishers are not making the right textbooks electronically available on the right terms.

When librarians were asked why they had not bought more e-books the main reasons were: e-books pricing models are not satisfactory (64%); there is too little choice of e-book titles (62%); e-book access models are not satisfactory (53%); we are waiting for the market to settle down (33%); and, we are waiting for JISC to offer better e-book deals (30%). The relatively poor take-up of e-books in academic libraries is further supported by the finding of the June 2007 Ebrary survey of 583 respondents from 67 countries (Ebrary, 2007). This showed that only 10% of libraries had over 50,000 e-books. However when asked whether there is any pressure on them to develop e-book collections 68% of librarians said 'yes'.

So what is it that librarians really want as far as e-books are concerned? The JISC report discovered the following:

- 'Reading list' materials, principally textbooks.
- Current titles.
- Wider choice relevant to the UK academic community.
- Flexibility in choosing between subscribing and outright purchase.
- Sensible charging bands or prices regime.
- Multiple and concurrent access for users.
- Easy access to the host server for Shibboleth/ATHENS users.
- A more systematic way of discovering what e-books are available.

From the perspective of the EBWG it was pleasing to note that the JISC sponsored report demonstrated that a high proportion of librarians felt that there were clear roles for JISC in the acquisition of e-books. 87% of librarians said that JISC performed a valuable national 'value-for-money' role seeking to get the best buys for the sector. 66% felt that JISC had an important role in investigating innovative formats or purchasing models, and 40% were of the opinion that JISC should facilitate the buying of resources that are essential in niche areas for teaching and research where libraries would not be able to afford the acquisition without financial help.

The findings of the report were carefully considered by the EBWG and following extensive discussions it was agreed that a top priority for the group should be firstly to devise a project to kick-start the e-book marketplace for libraries and secondly to focus activity upon core reading list materials - titles that individually librarians had had little success in acquiring. It was recognised that this was a difficult area as publishers are very wary about making textbooks or core readings available electronically – fearing the loss of student sales. With national funding it was hoped that the deadlock might be broken and both libraries and publishers could acquire real data that would inform future business models. And so the inspiration for the National E-Book Observatory Projectwas born.

The National E-Book Observatory Project

The EBWG was fortunate to acquire a considerable capital sum of money from JISC to fund the project. The aims of the project were threefold: firstly to license a collection of online core reading list materials that were highly relevant to UK HE taught course students in four discipline areas. The core reading list materials would be free at the point of use to all 180 HE libraries for a period of two years. Secondly the project would evaluate the use of the materials through deep log analysis and measure the impact of the 'free at the point of use' materials upon publisher, aggregator and library processes. Thirdly, knowledge acquired from the project would be transferred to all stakeholders to help stimulate an e-book market that has appropriate business and licensing models.

An invitation to tender (ITT) was sent to the publisher and aggregator communities inviting them to submit core reading list materials for inclusion in the project. The books could be in any of four subject areas: business studies, engineering, medicine (not mental health or nursing) and media studies. The ITT stated that £600,000 was available for the purchase of such materials and that it would go to multiple vendors and stringent criteria were stated about the actual texts offered. Materials must be for HE taught course students; they must be UK focused where relevant; and they must be made available on the platforms used by HE institutions. It was made clear that materials offered by publishers and aggregators would be evaluated by the education community and a priority list would be created against relevance, value for money and print sales - to verify demand.

A significant number of publishers and aggregators responded to the ITT and in June 2006 the Project Steering Group selected the bids that best matched the criteria to go forward for community consultation. In medicine e-book titles from Wolter Kluwer and Wiley Blackwell were selected; in engineering titles from Thomas Telford, Cambridge University Press and Elsevier; in media studies titles from Palgrave Macmillan and Taylor & Francis; and in business titles from Thomson Learning, John

Wiley, Cambridge University Press; Taylor & Francis, Palgrave Macmillan, Elsevier and Pearson. The cost of the e-book titles ranged enormously from around £5,000 to £184,000 for one title. However, it must be remembered that these titles will to be available (free of charge to individual libraries) to all 180 UK HE institutions for a period of two years and it was felt by JISC that publishers needed to be fairly compensated for the potential loss of income.

In addition to the cost of titles, a number of other issues were raised by the bids. The first related to e-book platforms. Prior to issuing the ITT, the EBWG had undertaken a survey of libraries about the most popular e-book platforms. It was hoped that in their bids to join the project, publishers would offer a range of platforms that their books could be accessed on. However this was not to be and most bids lacked platform flexibility and few bidders offered multiple platforms. It may be that the project will end up with books on several different platforms - which would be far from optimal. On a more positive note, the ITT stated that all publishers - if selected to join the project - must provide MARC records for their books for libraries to incorporate into their OPACs. The vast majority of publishers agreed to this condition. The Steering Group felt that this was important for a number of reasons, not least because the timescale for the project was very tight and this would save libraries from having to individually catalogue the books. In addition various surveys have shown that having OPAC records definitely drives up the use of e-books. The Ebrary survey, previously referred to, showed that the library catalogue was the most favoured route for users to find e-books.

In parallel with the issuing of the ITT to the publisher and aggregator communities for e-book content, a more general ITT was issued for the deep web log analysis. This tender was for a 12 month study - using raw logs of user transactions to assess the impacts and usage of the core reading materials. The data derived from such a study then generates questions for follow up web-based questionnaires which will be sent to staff and students in a range of libraries. Jointly the two methodologies will provide an understanding of the e-book behaviours of staff and students and will inform business models for e-books and strategies for embedding e-books into libraries. The results of the usage study will be shared with all participating libraries and publishers.

As stated earlier, the project timescale is tight due to conditions placed by JISC on spending the capital money. In July 2007, when this paper was presented to the LIBER Conference in Warsaw, the project had reached the point where e-book titles had gone out to community consultation. The timetable for the next stages of the project is as follows:

- August 2007 - bids awarded to a range of publishers/aggregators for core reading materials.
- July/August - launch activities (including marketing and promotional materials) to HE libraries in
- time for the start of the academic year in October 2007.
- October 2007 - start of the deep web log analysis for a period of one year.
- October/November 2007 & January/February 2008 - project road shows.
- October 2007 to December 2008 - embedding and evaluation.
- December 2008 - close of project (although e-book titles will be available to libraries until October 2009).

In summary it is felt that there will be significant advantages and benefits to librarians, publishers and aggregators from the National E-Book Observatory Project. The project will:

- provide an in-depth understanding of how e-books that support UK HE taught course students are
- actually used in teaching and learning;
- enable publishers, libraries and aggregators to assess the demand for core reading list e-books;

- enable all parties to measure the effect of 'free at the point of use' e-books on the buying behaviour
- of students;
- enable libraries to measure the benefits and potential costs of providing core reading list e-books to
- students;
- inform the creation of appropriate business and licensing models;
- inform the promotion of e-books within an institution;
- raise awareness generally of e-books throughout the academic community;
- stimulate the e-books market in a managed environment.

Further updates on the project will be available on the project wiki for those who wish to follow progress.

REFERENCES

1. CHEMS - The Higher Education Consultancy Group: A Feasibility Study on the Acquisition of e-Books by HE Libraries and the Role of JISC. Final report. October 2006. http://www.jisc.ac.uk/media/documents/jisc_collections/ebooks%20final%20report%205%20oct.doc

2. Ebrary. Global eBook Survey. Palo Alto : Ebrary, 2007. http://www.ebrary.com/corp/collateral/en/Survey/ebrary_eBook_survey_2007.pdf

9

CHAPTER

The e-Reader — an Educational or an Entertainment Tool? e-Readers in an Academic Setting

Peter Ahlroos, Jonna Hahto

Tritonia Academic Library, Vaasa, Finland

ABSTRACT

In this paper the authors will discuss a pilot project conducted at the Tritonia Academic Library, Vaasa, in Finland, from September 2010 until May 2011. The project was designed to investigate the application of e-readers in academic settings and to learn how teachers and students experience the use of e-readers in academic education. Four groups of students and one group of teachers used Kindle readers for varied periods of time in different courses. The course material and the textbooks were downloaded on the e-readers. The feedback from the participants was collected through questionnaires and teacher interviews. The results suggest that the e-reader is a future tool for learning, though some features need to be improved before e-readers can really enable efficient learning and researching.

Keywords: e-readers, higher education, academic libraries

INTRODUCTION

Aim

In the autumn of 2010 The Tritonia Academic Library started an e-reader project with funding allocated by the city of Vaasa, in honour of the city's 400-year anniversary. In 2010 there was a growing interest in e-readers and tablet computers, especially in the local press, which also spurred the idea to start a project to examine the use of portable reading devices in an academic setting. The purpose was also to gather more knowledge about e-books and e-readers in libraries and to promote the services of Tritonia.

The aim of the project is to investigate the application of e-readers in an academic setting and to learn how teachers and students experience the use of e-readers in academic education.

About Tritonia

Tritonia is a joint academic library for five universities: the University of Vaasa, VAMK University of Applied Sciences, Åbo Akademi University, Novia University of Applied Sciences and Hanken School of Economics. Tritonia also provides pedagogical services for the staff of all five universities. Of the 12,000 students, 70% have Finnish as their mother tongue, while the remaining 20% are Swedish-speaking.

The printed collections consisting of 336,000 volumes are available for all students and staff of the five universities and universities of applied sciences, as well as for external library users. By contrast, all electronic resources are licensed for a specific organisation and only the students and staff of the particular organisation have access to those. Because of this, one single e-book may have to be acquired separately for all the five organisations. This is challenging for both the library staff and the library users.

In addition to traditional library services, Tritonia also offers educational support at the EduLab unit. The activities that Edulab is based upon are pedagogical courses (25 ECTS), guidance and media production, which are offered for free to the staff of all five universities. Since the start in 2001, the main goal has been staff development and support.

Background

The e-book and e-reader market is booming at the moment. Amazon already announced that they are selling more Kindle books than print books and Apple's iPad dominates the tablet computer market (Amazon.com, 19.4.2011). In Europe the e-book market has evolved at a slower pace, mainly because of the lack of content and the high price of e-readers and e-books, but the Kindle's arrival to Europe has livened up the market.

Libraries have also invested in e-books, have purchased e-book databases, have tried out lending e-readers to customers and set up pilot projects. Even so, despite of their best efforts, the libraries have not really forged their way into the e-book business. However, for the future it is more vital than ever that libraries continue to experiment with different e-book models and e-book and e-reader solutions.

Research in the Same field

Several studies concentrating on e-books and articles about e-readers were published at the beginning of the 21st century. However, the devices never really broke through until the past couple of years. At the moment more and more studies again focus on e-readers, e.g., Clark et al., (2008) dealt with user experiences and users' opinions about e-readers in their survey conducted at Texas A & M University Libraries. The participants in the study, members of the library and university staff, found the Kindle reader interesting, but questioned the potential usefulness of the device in research libraries on account of licensing issues, graphic display and cost.

Pattuelli and Rabina (2010), on the other hand, attempted to understand the effects of reading devices on everyday reading practices and also potential applications for library services. They gathered feedback from 20 library and information science students, who had used the devices for one week. The results were mixed; however, the usability issues were outweighed by the device's portability and readers' convenience of use anywhere.

Both Clark et al., (2008) and Pattuelli and Rabina (2010) focused mostly on leisure reading. The use of e-readers in an academic setting and the potential application of the e-reader as an educational tool have been much less explored themes ine-reader studies. There are a few exceptions, including pilot studies conducted by seven universities in the USA (Damast, 2010) and Aalto University in Finland (Aaltonen et al., 2010). In both studies the students were given e-readers with course material for a certain period of time. The studies encountered certain usability issues. For example, students felt that the lack of colours and the difficulties in navigation, browsing, bookmarking and taking notes made the devices awkward to use.

An Academic Setting — a Challenge to e-Readers

The purposes for reading and the methods used can offer a starting point for a discussion of the challenges for e-readers, because the way we read varies with the type of literature we read (see Marshall, 2010, p. 33). For example, reading a fiction novel differs substantially from reading a textbook, as in leisure reading the text usually advances linearly. Reading for study and research purposes involves a lot of browsing, glancing, seeking and re-reading. Annotation and navigation are important ways to interact with the text (Marshall, 2010, pp. 37–36.). In order for e-readers to be considered a real alternative or replacement for paper books and computers, reading devices should be able to support different ways of using the devices.

Today's reading devices are mainly focused on linear reading or, more specifically, reading for entertainment. Most of the devices on the market have annotation and search functions, but as studies indicate, users often find these functions awkward and slow to use (Aaltonen et al., 2010; Damast, 2010). Also, browsing books is often found difficult. Another possible weakness emphasized in an academic setting is the lack of colours, as pictures, charts and tables are important parts of scientific documents.

The different formats can also create difficulties for students and researchers. e-Books in ePUB format, for example, do not have static page numbers, which makes it very hard to provide precise page numbers for citations. PDFs do not have the same problem, but because the PDF format is not scalable, books and documents in this format may require scrolling and are therefore challenging to read on some e-readers (Ballhaus et al., 2010, p. 3).

Teaching, Learning and e-Books

The significance of information technology in teaching cannot be underrated as the students are already familiar with information technology both in informal and formal settings. They are part of the Generation Y, the Net Generation, accustomed to navigate in the constant information flow. However, the filtering strategies they use to cope with large quantities of information can often be described as hasty and superficial, which reflects how they regard and use information (Parment, 2008). This generation is also more demanding than the previous one, expecting teachers and librarians to keep up with the new technology.

As Cox (2010) points out, the teacher's attitudes and beliefs as well as the teacher's willingness to adopt new teaching methods and practices affect the perceived value of IT in learning. Unfortunately, studies indicate that once their training is finished, teachers rarely see the need to question or change their teaching practices, including those related to the use of IT (Cox, 2010, pp. 16–17) A number of researchers have looked at the influence of new technologies on student achievement and learning processes, but the results are both for and against (Minčić-Obradović, 2011, pp. 174–175). For one thing, the potential change tends to be temporary and superficial and teaching and learning methods are seldom transformed profoundly.

e-Books have the potential of being an efficient and effective educational tool, as the combination of text and multimedia can enhance readers' understanding of abstract concepts and theories (Ballhaus et al., 2010, p. 13; Collier and Berg, 2011, p. 32; Minčić-Obradović, 2011, p. 161). Features such as customisability of text size, the ability to convert text to audio and a built-in encyclopedia also have benefits in education (Minčić-Obradović, 2011, p. 161). Textbooks read on e-readers enable real-time dialogue and discussion between teachers and students, with options for sharing notes, comments and questions. Many e-readers do not yet have these features, but technical development has been so rapid that no doubt we get to see those in the future.

METHODOLOGY

The project was designed to investigate the application of e-readers in an academic setting and to learn how teachers and students make use of e-readers during courses. The data was mainly collected through questionnaires and interviews. In the next paragraphs, we will discuss the choice of e-reader, preparing the test groups, and the data collection methods.

The main reasons for choosing Amazon's Kindle reader were: a user-friendly interface, an affordable price, a range of functions such as dictionary and text-to-speech and the possibility to download one e-book onto several devices. The closed environment, the lack of EPUB support and the lack of Finnish or Swedish content were not considered major issues, as the aim of the project was not to concentrate on usability issues or to evaluate the device itself, but to focus on the pedagogical aspect of e-readers in general.

The e-reader project started in September 2010, once the third-generation Kindle became available, and ended in May 2011. During the project five different groups of students experimented with e-readers. The groups are presented in Figure 1.

Subject and course	Number of students in the course	Period of use	Books
Technical Communication, research course in advanced studies	10	4 months	George F. Hayhoe: A Research Primer for Technical Communication
Journalistik, Mass media and communication	9	4 weeks	Joseph Turow: Media Today (3rd Edition)
Information Technology, database management in advanced professional studies	6	4 weeks	M. Lisin, J. Joseph and A. Goyal: Microsoft SQL Server 2008 Reporting Services Unleashed
Language (Swedish), Digital texts	11	4 weeks	Books from project Gutenberg, Maria Turtschaninoff's fantasynovel 'Underfors'
Learning in Networks 1, teacher group	15	3 weeks	Confronting the Challenges of Participatory Culture, Living and Learning with New Media

Figure. 1 Test groups.

The groups chosen for the project were small, advanced courses, where the teacher was able to choose the literature used in the course. We approached specific teachers because we knew them to be interested in trying out new methods and techniques. The books and other documents chosen by the teachers were downloaded on 25 Kindle readers (wifi). Later on five additional Kindles were acquired. Kindle DX, Sony PRS-650 reader touch edition, Bokeen Cybook OPUS and iPad I were also acquired, mainly for demonstration purposes.

A short introduction (20–30 minutes) was arranged for the test groups and the students were asked to sign a lending agreement. The participants were also given the possibility to download their own documents on the Kindle reader. The introduction was the only contact with the students and further communication was handled by the teacher.

The students were asked to answer online questionnaires before and after the test period. Due to the project timing and resourcing, the only manageable way of collecting data was through a questionnaire. The questionnaires included open questions as well as multiple choice questions. In the first questionnaire the respondents were asked about their technical skills, studying habits and attitudes towards new technology in general. The second questionnaire related to studying in the course, the use of the device and respondents' thoughts about e-readers, studying and learning. Three teachers were interviewed to get in-depth insight into students' reactions and teachers' thoughts about e-reader applications. We also had access to a blog article and six reviews written by the students as additional assignments in their course.

In addition to the test groups, e-reader and e-book theme days were organised. During these three days over 200 students, staff members and other visitors acquainted themselves with different e-readers and the library's e-book collections. 67 visitors also filled in a questionnaire, which contained questions about the library's e-book collections and attitudes towards e-readers.

RESULTS

89% of the students filled in the first questionnaire but only 56% (20 of 32) of the students answered the second questionnaire, despite of several reminders sent by email. In the teacher group 10 of 15 teachers filled in the questionnaire. Both the students and the teacher group can be categorised as early adopters or as an early majority on the innovation adoption curve of Rogers (2003, 281), as shown in the graph below. The majority of the respondents estimated their technical skills above average.

In the next sections the respondents' attitudes towards the Kindle reader and Kindle's pedagogical value will be discussed.

Respondents' Attitudes towards the Kindle Reader

One of the findings of the theme days and the student group meetings was that the e-readers seem to lack the wow-factor, which makes tablet computers like the iPad so attractive. As a couple of respondents pointed out, the lack of colours, basic button navigation and poor internet browsing capabilities made them think of 'old black-and-white TVs' and 'the Middle Ages'. Still, half of the respondents said they would rather read documents on a basic e-reader than on a computer screen.

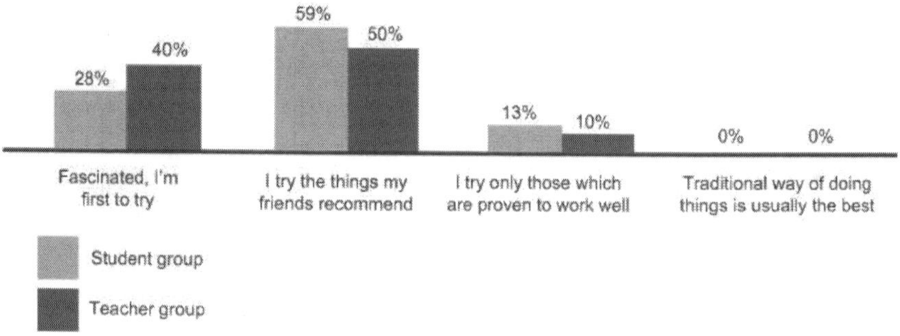

Figure. 2 Respondents' attitude towards new technologies.

The teacher group appeared to be slightly more critical towards the Kindle reader than the students, especially judging by the answers to the open questions. In many comments the Kindle reader was compared to the iPad and found to be inadequate, as 'Kindle cannot compete with iPad, there's so much more you can do with an iPad'. Even though the students in general gave lower values for different features, they seemed to be more neutral in their attitudes towards the Kindle reader. Many students did not seem to care whether they read the book in paper format or on an e-reader as long as they had access to the book. Based on the students' comments one may speculate that changing the reading habits and 'getting used to the different format', as several students put it, is really a bigger issue for the students than the shortcomings in usability.

The teacher interviews also gave clues about the way the students experienced the use of the e-readers, indicating that for them it would not necessarily be a problem to have a device for one purpose only. The students seem to want to keep their studies and free time apart, and for this reason it is natural for many of them to have different devices for different purposes. That is, the iPad is for 'all the fun stuff' like surfing, whereas the Kindle reader is a more serious device, meant for studying and reading. One of the students actually noted that it is a good thing that the Kindle reader doesnot have any unnecessary functions, because it is easier to concentrate on studying.

One of the teachers observed that the students seemed to take the course more seriously and that they were more active and ambitious than usual, but it is impossible to determine if this was a reaction to the new, interesting technology or if there simply were more enthusiastic students in the group than usual.

In Figure 3, the overall user experience is shown. Nearly every respondent had a positive or at least a neutral experience with the Kindle reader. A majority of both the teachers and the students also estimated that the effect of the e-reader on studying and learning on the course was neutral or slightly positive.

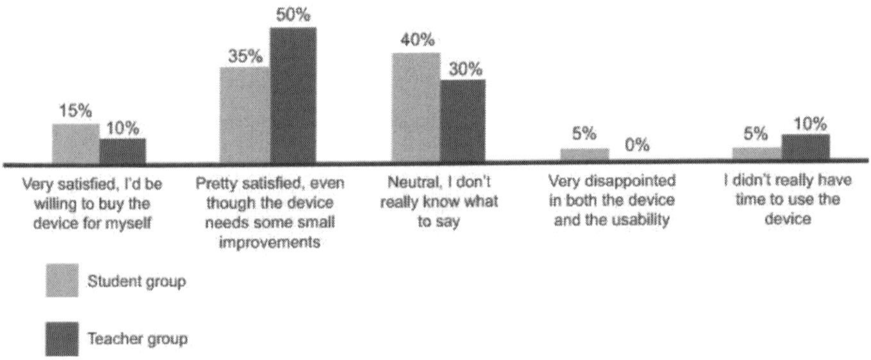

Figure. 3 Overall user experience.

As assumed by previous studies (Aaltonen et al., 2010; Ballhaus et al., 2010; Damast, 2010), the features the respondents missed the most were a touch screen, colours, internet browsing, real page numbers and scandic letters. Many of the teachers were especially dissatisfied with Kindle's PDF support, as reading the PDF file requires a lot of navigation and the size of the text is small. The most appreciated features were the readability of the text, the portability, and the long battery life of an e-reader.

The Kindle Reader in Learning and Teaching

Only a few respondents had previous experience with e-readers, so they — as was expected — found it difficult to evaluate the potential application of the device in teaching and learning. The students were most pleased that they did not need to chase textbooks in the library and carry heavy book bags to school. The three teacher interviews gave some insight into possible ways of using e-readers.

Teacher 1 found it very important that his students, as future journalists, would read and explore text in the same way as their future readers will explore the text written by the students. The use of the e-reader caused the teacher to rethink his teaching methods and also to present a concrete plan as to how the course could be developed utilising the Kindle reader to enhance the students' understanding of the subject. He also had ideas about using the Kindle for presenting reports.

Teacher 2 held this course for the first time and therefore did not yet have any ideas for renewing it. The idea of the course was to explore digital texts and the teacher got special permission from the publisher to use a Swedish fantasy novel as course material. The students were satisfied that they could read a book without others knowing what they read and that they could alter the font size. As future primary school teachers, they also speculated whether the e-reader could increase the motivation for reading among boys. These students felt that reading fiction would enhance the experience with multimedia and linking to web pages. Especially the fantasy novel with many references to real places would have given them a better reading experience if they could have used the web while reading.

The third interviewed teacher was not fully convinced of the benefits of e-readers in learning and teaching, mainly because of the difficulties in browsing books and the poor support for PDF files. In her courses books are usually used as reference literature and with the Kindle it was hard to navigate between chapters and get an overview of the book. She also uses lots of PDF articles, which are awkward to read on the Kindle since they require scrolling.

In her opinion, the ideal teaching aid would not be the traditional book, but a service where a selection of different devices would be offered together with digital materials and pedagogical support. Another function she desired was a shared notation system.

DISCUSSION AND CONCLUSIONS

Even though the study was a pilot project and the results may not be generalised because of the small and homogeneous sample, the feedback has been encouraging and we intend to continue the project in the autumn. There are features such as browsing, PDF support and an internet connection, that need to improve before e-readers can enable efficient learning and researching in an academic setting. However, though many of the respondents still preferred the paper book, nearly all of them saw the e-reader as a future tool for studying.

As expected, those teachers with sparse experience of either the Kindle, the content of a specific e-book or the course, had few ideas on how to develop their courses. One month of e-reader usage seems to be too short a time period and the teacher test period should really start much earlier than the course. Consequently, in the future the project will focus on teachers and they will be offered a longer loan period.

The vision for the future of libraries is that students and teachers would be able to download library material easily from a distance on their own devices. For the time being e-books or e-readers are not a solution, especially because the vast majority of the libraries' present e-book collections are not compatible with most of the e-readers, and the books read on e-readers are often available only for private customers. Logistic issues such as charging the batteries and preparing the e-readers also require a great deal of time and effort from the library staff, if the reading devices are lent to library patrons. However, if libraries want to become part of the development and the chain of distributing e-books, it is essential that they experiment with various e-book distribution models and e-readers.

The volume of e-resources increases every year and remote access to resources reduces the need to visit the library. This does not necessarily mean that libraries are less important than before, it merely represents a change in patrons' needs. As the collections are more and more in digital format and the self-service increases, libraries have more resources to concentrate on the quality of service rather than the quantity. There is a greater need for information services, personal guidance and information literacy education than ever.

Finally, from our point of view the project can be considered a success merely because it got some of the teachers to rethink their methods of teaching and ways to develop the course. As a whole, the project shows that academic libraries can contribute to teaching methods by providing tools and support to teachers to develop their teaching methods and acquire new ideas and technology.

REFERENCES

1. Aaltonen, M., P. Mannonen, S. Nieminen and M. Hjelt (2010): 'A collaborative study: on the demands of mobile technology on virtual collection development',World Library and Information Congress: 76th IFLA General Conference and Assembly. 10–15 August 2010, Gothenburg, Sweden. http://www.ifla.org/files/hq/papers/ifla76/151-aaltonen-en.pdf.

2. Amazon.com, news release 19.4.2011: 'Amazon.com Now Selling More Kindle Books Than Print Books'. http://phx.corporate-ir.net/phoenix.zhtml?c=176060&p=irol-newsArticle&ID=1565581&highlight=.

3. Ballhaus, W., M. van der Donk and P. Stokes (2010): 'Turning the pages: the future of ebooks', Technology, Media & Telecommunications, pwc. http://www.pwc.com/gx/en/entertainment-media/publications/future-of-ebooks.jhtml.

4. Clark, D. T., S. P. Goodwin, T. Samuelson, and C. Coker (2008): 'A qualitative assessment of the Kindle e-book reader: results

from initial focus groups'.Performance Measurement and Metrics, 9(2), 118–129.

5. Collier, J. and S. Berg (2011): Student learning and e-books. In: S. Polanka (ed.), No shelf required: e-books in libraries (pp. 19–36). Chicago: American Library Association.

6. Cox, M. (2010): The changing nature of researching information technology in education. In: A. McDougall (ed.), Researching IT in education: theory, practice and future directions (pp. 11–24). New York, NY: Routledge.

7. Damast, A. (2010): 'E-Book Readers Bomb on College Campuses'. BusinessWeek(June 11, 2010), 3.

8. Marshall, C. C. (2010): Reading and Writing the Electronic Book. Synthesis Lectures on Information Concepts, Retrieval, and Services #9, San Rafael, California: Morgan & Claypool Publishers.

9. Minčić-Obradović, K. (2011): E-books in academic libraries. Oxford, UK: Chandos.

10. Parment, A. (2008): Generation Y: framtidens konsumenter och medarbetare gör entré. Malmö: Liber.

11. Pattuelli, C. M. and D. Rabina (2010): 'Forms, effects, function: LIS students' attitudes towards portable e-book readers'. Aslib Proceedings 62(3), 228–244.

12. Rogers, E. M. (2003): Diffusion of innovations. 5th ed. New York: Free Press.

10

CHAPTER

Knowledge and Skills for the Digital ERA Academic Library

J. Raju

Library and Information Studies Centre, University of Cape Town, Private Bag X3, Rondebosch, 7701 South Africa

ABSTRACT

Technology has altered the traditional academic library beyond recognition. These dramatic changes have impacted significantly on the knowledge and skills requirements for LIS professionals practising in this environment. While there have been studies in other parts of the world which have investigated the knowledge and skills requirements for the digital era academic library environment, to date no comprehensive study has 'drilled' down into this area in the South African context. This paper reports on a preliminary study which is part of a wider study aimed at developing a comprehensive skills statement which would provide an objective framework against which professional LIS practitioners in the modern academic library environment in South Africa may both measure their existing competencies and also identify the need for further skills acquisition. The research question guiding this preliminary investigation was: What key knowledge and skills are required for LIS professionals to effectively and efficiently practise in a digital era academic library in South Africa? The triangulated findings (using content analysis of job advertisements and semi-structured interviews) from this preliminary investigation are used to ascertain an initial picture of key knowledge and skills sets required for LIS professionals in this environment. These preliminary findings also proved useful in teasing out some of the parameters for the wider study targeting the development of a comprehensive skills statement for higher education libraries in South Africa. The study reported here has relevance for the academic library context in other parts of the world as well.

Keywords: Academic libraries; Digital era; Disciplinary knowledge; Generic skills; Job advertisements; Personal competencies

INTRODUCTION

New methods of scholarly communication, expansion of the library's virtual space via knowledge or research commons, the proliferation of social media, and the explosive growth of mobile devices, tablets and related applications, have collectively altered the traditional academic library beyond recognition. These dramatic changes, largely the result of rapidly evolving information and communication technologies (ICTs), have impacted significantly on the knowledge and skills requirements for library and information science (LIS) professionals practising in this environment. The transformed landscape requires a new generation of LIS professionals to effectively and efficiently mediate it. Orme (2008, p. 627–628) categorised knowledge and skills required for this transformed environment into: discipline-specific knowledge (that is, knowledge that relates specifically to the LIS profession), generic skills (general skills which apply to all disciplines) and personal competencies (attitudes, values and personal traits). Choi and Rasmussen (2009, p. 465), through content analysis of job advertisements in the United States of America (USA), found that key disciplinary knowledge required for this digitally oriented environment included understanding metadata, and knowledge and experience in digital content creation and management. Generic skills such as effective communication and interpersonal skills, critical thinking, problem solving and teamwork were found by Nonthacumjane (2011, p. 283) to be required by information professionals in a digital library environment in both Norway and Thailand. Howard's (2009) Australian master's study (cited in Nonthacumjane, 2011, p. 283) highlighted personal competencies such as flexibility, adaptability and reflective thinking as being required for working in a digital library environment. What about knowledge and skills requirements for the digital era South African higher education academic library environment? To date, no comprehensive study has 'drilled' down into this area in the South African context. South Africa is richly endowed with academic libraries. These are located in the country's 23 universities. Many of these higher education libraries offer state-of-the art LIS services to academic and research communities in universities in the country which are leading institutions on the African continent. Hence an analysis of knowledge and skills requirements for the digital era academic library environment in South Africa is important.

This paper reports on a preliminary study which is part of a wider study aimed at developing a comprehensive skills statement which would provide an objective framework against which professional LIS practitioners in the modern academic library environment in South Africa may both measure their existing competencies and also identify the need for further skills acquisition. Such a skills statement would also be useful in informing curriculum review and revision in LIS education and training, as academic libraries in South Africa are a major employer of LIS graduates (Ocholla & Shongwe, 2013, p. 38). There are ten LIS schools in South Africa, based in universities, which offer education and training in response to the knowledge and skills requirements of the LIS work environment. The research question guiding this preliminary investigation was: What key knowledge and skills are required for LIS professionals to effectively and efficiently practise in a digital era academic library in South Africa? The triangulated findings (using content analysis of job advertisements and semi-structured interviews) from this preliminary investigation are used to ascertain an initial picture of key knowledge and skills sets required for LIS professionals in this environment. These preliminary findings also proved useful in teasing out some of the parameters for the wider study targeting the development of a comprehensive skills statement for higher education libraries in South Africa. While these early findings draw from South African higher education libraries, the study is contextualized within global trends affecting higher education libraries and hence has relevance for other parts of the world as well.

LITERATURE REVIEW

A review of pertinent literature locates this study in the global context of the changing academic library and its impact on knowledge and skills requirements for academic libraries in the digital era.

The academic library in the digital era

As technology continues to impact on the delivery of information services, traditional academic library systems, point out Choi and Rasmussen, 2006 and Choi and Rasmussen, 2009, have come to embrace the digital library model. Academic libraries have evolved from focusing on the management of physical resources and related services to "transforming resources and services into digital formats to support teaching, learning and research" (Choi & Rasmussen, 2009, p. 457). O'Connor and Au (2008, p. 57) too state that the academic library has changed dramatically and with this change the "conception of a digital library has become a reality".

Changes in areas of teaching and learning, influenced and enabled by technology, remarks McCarthy (2005), have impacted on academic libraries — for example, the creation of new knowledge products such as subject portals and subject specific websites to support teaching and learning; or the re-purposing of physical spaces and the expansion of virtual spaces to support new pedagogies and changes in the teaching and learning process.

In terms of research, Luce (2008) explains that the "convergence of exponential increases in computing, storage, online sensors, and bandwidth" has enabled scientists and researchers to collaborate in new ways thus leading to the rise of eScience and eResearch. eScience developments now characterise not only the Science, Technology, Engineering and Medicine disciplines but the Humanities and Social Sciences as well. The convergence of technologies has led to "new ways of thinking about and understanding physical, biological and social phenomena" (Luce, 2008). These revolutionary developments have demanded an equally dramatic shift in the way academic or research libraries serve the needs of scientists in the new eScience and eResearch contexts.

Luce's (2008) advice to academic libraries which are positioning themselves to support eResearch, is to be cognizant of the fact that knowledge preservation becomes one of the key roles of such a library. Luce (2008) goes on to emphasize that this knowledge preservation necessitates:

- Ensuring the quality, integrity, and curation of digital research information;
- Sustaining today's evolving digital service environments;
- Bridging and connecting different worlds, disciplines, and paradigms for knowing and understanding; and
- Archiving research data in a data world.

Hence to enable digital capture, curation, preservation and sharing of knowledge, the academic library in the digital era needs to reflect a service environment that embraces digitization, electronic publishing, Web 2.0, Web 3.0, Library 2.0, Library 3.0, social media, open access, and a host of other fast evolving ICTs. As academic libraries the world over shift into this digital era, these developments and innovations impact on the knowledge and skills profiles of LIS professionals in academic libraries (Choi and Rasmussen, 2009 and Nonthacumjane, 2011). New skills sets are required to mediate this digitally oriented academic library environment.

Knowledge and skills requirements in the digital academic library environment

A study of job advertisements by Orme (2008, p. 630) found that "a mixture" of discipline-specific knowledge (also referred to as professional knowledge), generic skills and personal competencies is required of LIS professionals in a digitally oriented LIS environment. Nonthacumjane too (2011, p. 280, p. 286), in a study that probed the "skills and competencies required for LIS professionals to be effective and efficient working in the digital era", concludes that they should be able to dynamically exercise "personal, generic and discipline-specific skills".

Disciplinary knowledge

Partridge and Hallam (2004) use the "double helix image of human DNA" to argue that both disciplinary knowledge and generic capabilities "make up the genome of the successful information professional in the information age". Many have argued that the new digitally oriented academic library is a renewed conceptualisation of traditional LIS resources and services, now undergirded and driven by new technologies (Choi and Rasmussen, 2009, Gerolimos and Konsta, 2008, McCarthy, 2005, Middleton, 2003 and Missingham, 2006). Hence in a digital academic library environment core knowledge and skills of traditional librarianship are important but need to be augmented by new technological knowledge (Choi & Rasmussen, 2009, p. 465).

Tammaro (2007, p. 237) cogently points out that

cataloguing and classification skills have much relevance to the Web...a more thorough knowledge of the major schemes and their working principles is required to allow a person to adapt and accommodate existing metadata schemes to use, and to possess the basic expertise to construct new schemes.

Digital library applications are closely linked to Web technology (Choi & Rasmussen, 2009, p. 463). Consequently, as modern academic libraries move into the creation of digital content and its organisation and preservation through metadata creation and management to make their special collections more accessible via the Web, the need for knowledge of the following technologies becomes critical: digital library architecture and software, technical and quality standards, HTML coding, general computer skills and computer literacy, database development and management, Web mark-up languages such as SGML and XML, and Web development and design (Choi and Rasmussen, 2006 and Choi and Rasmussen, 2009). Ocholla and Shongwe (2013, p. 39, 42) in their content analysis of job advertisements in South Africa over a four-year period (2009–2012) found IT skills to be very sought after in libraries as "more information services", particularly in academic and research environments, become "IT or e-access and e-service dependent".

Parallel to the development of digitization and curation of unique collections, the academic library in the digital era is also being challenged by an emerging trend of research data management and curation (Wise, Henninger, & Kennan, 2011, p. 279). The emergence of eScience and eResearch is accompanied by the generation of vast amounts of research data in need of collection, preservation, management and provision for future access to enable re-using, re-purposing, re-combining, etc. According to Luce (2008) metadata is an essential component of research data and research or academic libraries, because of their traditional knowledge base, are well positioned to "lead the development of standardized, ontologically rich automated metadata" for research data sets. Here again, disciplinary knowledge, that is, the creation and management of metadata which are established tasks in the LIS profession, allows academic libraries to take responsibility for the curation and preservation of data for its re-use when needed (Luce, 2008). The Ocholla and Shongwe (2013, p. 41) study makes reference to "new job titles" emerging in the LIS job market in South Africa and puts forward that "these new titles represent strong ICT elements" and shows the influence of ICTs on emerging knowledge and skills requirements in the LIS sector. In the USA, Riley-Huff and Rholes (2011, p. 129–130, p. 135) make a similar observation in their survey of library administrators and librarians from academic libraries: "new job categories are being defined in LIS" which "while centering on core librarianship principles [knowledge organisation; knowledge dissemination; etc.]", or what this paper has been referring to as disciplinary knowledge, call for a "significant technology skills set" (a point alluded to earlier).

Generic skills

Also referred to as 'transferable skills' or 'graduate attributes' (Partridge & Hallam, 2004), generic skills refer to life skills such as communication and interpersonal skills, critical thinking, problem solving and teamwork which allow individuals to function not only in disciplinary or subject domains but also in employment and social situations. Orme's (2008, p. 626) content analysis of job advertisements in the USA revealed that from among disciplinary, generic and personal skills, "generic skills were most frequently sought", followed by disciplinary and then personal skills. Orme (2008, p. 626), however, emphasizes that while this supports the 'move to the generic' argument in the literature (Kennan et al., 2006 and Wise et al., 2011) (that is, it has become important to appoint individuals with, for example, a capacity for continuous learning and who are adaptable in a fast changing work environment rather than those who simply possess specific disciplinary knowledge), disciplinary knowledge remains important to LIS employers. This, according to Orme (2008, p. 626), is evidenced by the marked presence of disciplinary or professional skills in the top twenty most frequently sought requirements examined.

The literature reveals communication skills to be among the most highly ranked generic skills (Gerolimos and Konsta, 2008, Middleton, 2003, Orme, 2008 and Reeves and Hahn, 2010). Of particular relevance to the modern academic library is the observation that the demand for interpersonal skills is a reflection of the "reality that team-based approaches are a common practice in digital projects and interpersonal skills are a key to success in team efforts" (Choi & Rasmussen, 2009, p. 464). While general computing or computer literacy, like information literacy, would categorise as a generic skill, technology skills, for the purpose of this paper, have been discussed above under disciplinary skills because of the close alignment between digital library applications and ICTs (Riley-Huff & Rholes, 2011, p. 130). A similar principle is applicable to managerial skills, particularly in the context of the management of digitization, curation and data management projects.

Personal competencies

Choi and Rasmussen (2009, p. 457) identify from the literature the following personal attributes as being important in the LIS work environment: capacity for continuous learning, flexibility, fostering change and the capacity to work independently. To this list may be added enthusiasm and self-motivation (Orme, 2008, p. 621–622), reflective thinking, and the ability to respond to others' needs (Howard (2009) cited inNonthacumjane, 2011, p. 283). Choi and Rasmussen (2009, p. 464) highlight "adaptive skills to keep up with changes and challenges within library and information environments".

While some studies, for example Partridge and Hallam (2004), have conflated generic skills and personal attributes into a single category of generic capabilities, this paper has kept them separate for ease of analysis of data collected. It is evident from the review of literature that findings from studies of job requirements in the LIS professional sector, including the academic library sector, demonstrate the need for new generation LIS professionals to be "multi-skilled" (Orme, 2008, Reeves and Hahn, 2010 and Wise et al., 2011). In order for LIS professionals to effectively mediate the digital academic library environment, they would need to embrace a blend of discipline-specific knowledge, generic skills and personal competencies. Partridge and Hallam (2004) reiterate this: "...generic capabilities [which may include personal competencies] and discipline knowledge are quite significantly intertwined and interrelated and vital for success as a library and information professional in the twenty first century". Orme's (2008, p. 629–630) study concludes that to this "mixture" one needs to add experience as a significant number of job advertisements require experience in disciplinary and/or generic skills.Reeves and Hahn (2010, p. 115) too, in their survey of job advertisements, found that "a great many job ads requested work experience in specific functional areas of libraries...", particularly in technical service functional areas such as that requiring "cataloguing and metadata experience".

Table 1 summarises the preponderance in the literature of some of the broad trends relating to knowledge and skills requirements in the digital era academic library environment. In attempting such a table of findings from the literature, the researcher selected for inclusion in the table, only literature that:

- reported more recent studies (roughly the last 10 years, 2004–2013, but favouring more recent years);
- reported on key survey studies involving content analysis of mainly job advertisements (as the present study does) as well as

literature and practitioner/employer surveys on LIS knowledge and skills requirements; and,

- included at least one recent significant South African study in the area of LIS workplace knowledge and skills requirements.

Table 1. Knowledge and skills requirements for the digital era academic library.

Knowledge & skills requirements	Partridge and Hallam (2004)	Kennan et al. (2006)	Missingham (2006)	Choi and Rasmussen (2006, 2009)	Gerolimos and Konsta (2008)	Orme (2008)	Reeves and Hahn (2010)	Nonthacumjane (2011)	Riley-Huff and Rholes (2011)	Wise et al. (2011)	Ocholla and Shongwe (2013)	%
Mixture of disciplinary knowledge, generic skills & personal competencies	√					√	√	√		√	√	55%
Disciplinary knowledge still has a strong presence in the list of sought after skills	√			√		√	√	√				45%
Generic skills are being increasingly emphasized by employers as priority skills requirements		√		√	√	√	√			√		55%
Personal skills are considered important but lag behind disciplinary knowledge and generic skills in priority				√		√		√				27%
Strong demand for technology skills associated with LIS applications in the digital era			√	√	√		√	√	√		√	64%

While '...technology associated with LIS application...' is treated as a separate trend inTable 1 (for purposes of honing in on this significant trend), in other parts of this paper, as explained earlier, it is discussed under disciplinary skills due to the close association between digital library applications and ICTs. Table 1 also provides percentages (calculated against the 11 core research papers selected by the researcher for inclusion in the table) reflecting relative preponderance in the literature of the identified trends.Table 1 proves useful, in the Discussion section of this paper, in effecting comparisons between the findings of the preliminary study being reported and that in the literature reviewed in this paper.

METHODOLOGY

The preliminary study reported here adopts a qualitative approach by using content analysis of job advertisements and semi-structured interviews with purposively selected academic library professional LIS personnel to preliminarily ascertain the knowledge and skills requirements for a modern academic library in South Africa. As mentioned at the outset, the research question guiding this preliminary investigation was: What key knowledge and skills are required for LIS professionals to effectively and efficiently practise in a digital era academic library in South Africa? Content analysis is a commonly used descriptive technique for analysing the content of a document or other communication to discover features and patterns (Neuman, 2006, p. 44). Content analysis of job advertisements is a well-established method of researching requirements of the employment market in a particular sector (Orme, 2008, p. 620, p. 623) and is very useful in reflecting the demands of employers, employment opportunities and emerging trends in the employment market.

In view of the fact that this study was a precursor to an envisaged wider study, the *Mail & Guardian*, a weekly newspaper known nationally for carrying advertisements for vacancies in the higher education sector in South Africa, was initially used as a single source of advertisements for professional LIS posts (that is, posts requiring a professional LIS qualification) in academic libraries in South Africa. The researcher (towards the end of 2012) attempted to source relevant advertisements from the popular 'Jobs and Academic' section of the newspaper made available to the researcher at the newspaper's Cape Town satellite office. The researcher began searching, retrospectively, for advertisements from 2008 onwards on the assumption that this would deliver a significant number of 'recent' advertisements for LIS professional positions in academic libraries.

The *Mail & Guardian* did not have a publically available electronic means of searching its content — hence the retrospective searching had to be done manually. The researcher did not have access to issues of the newspaper that were at the binders at the time of searching. Added to this was the fact that a more recent trend is that many institutions do not place the full advertisement in the press — often an abbreviated form, largely for reasons of costs, appears with a URL to access the full advertisement which is removed from the relevant website once the closing date has passed, thus 'blocking' access to important research data. Reeves and Hahn (2010, p, 118) also allude to this difficulty: "... job ads typically appear now online, ...a distinct advantage for job-seekers, but a serious problem for future studies of this type if the ads are not captured and stored before they disappear". The net result of the retrospective searching for advertisements was that only 39 were found. The researcher supplemented this by searching differently for 2013 advertisements. The *Mail & Guardian* was purchased on a weekly basis and scanned immediately for relevant advertisements, including searching websites referred to by abbreviated advertisements. Over and above this, *LiasaOnline*, which is the listserv of the Library and Information Association of South Africa (LIASA), was accessed for relevant advertisements, following the advice that "...online job ads are essential to getting a comprehensive picture of the job market (Reeves & Hahn, 2010, p. 105)". This approach for the whole of 2013 yielded a further 32 advertisements bringing the total to 71 (excluding duplicate) advertisements.

To triangulate findings from the job advertisements for this exploratory exercise, the researcher purposively selected, for conducting semi-structured interviews, three professional LIS staff members each (six interviews in total) from the academic libraries of the University of Cape Town and Stellenbosch University, two of South Africa's research-led universities with world rankings — an obvious indicator of their state-of-the art academic libraries. It was also the intention of the researcher to use this modest number of interviews to trial this particular method of data collection for this type of study.

The interviewees in the purposive sampling were selected based on the researcher's knowledge and research experience (Welman, Kruger, & Mitchell, 2005, p. 69) with the identified research sites (the two academic libraries). This purposive selection included individuals occupying positions such as Senior Librarians (2) of branch library collections (Medicine and Law), Manager (1) of a disciplinary information service (Commerce), an institutional repository Manager (1), the Head (1) of Digitization Services and a Director (1) of Client Services and e-Scholarship in the university library and information services. The researcher was confident that the occupants of these positions would be a rich source of data needed to respond to the research question generated for this preliminary investigation. Each interview lasted about an hour and was based on a set of questions around disciplinary knowledge, generic skills and personal competencies relating to the interviewee's position as well as the LIS service in general. Depending on the responses from the interviewees, the researcher, where possible, probed further in an attempt to enhance the richness of the data collected.

ANALYSIS AND FINDINGS FROM THE PRELIMINARY INVESTIGATION

This preliminary investigation did not focus on job titles and qualifications required but rather scanned the contents of advertisements for requirements relating to disciplinary knowledge, generic skills and personal competencies. These knowledge/skills categories were identified in the literature (Choi and Rasmussen, 2009, Nonthacumjane, 2011 and Orme, 2008) reviewed for this study as being the main categories into which the knowledge and skills requirements for LIS professionals in a digital LIS environment, may be grouped. A chart (see Table 2) was created with these categories and each of these categories was populated with relevant knowledge and skills that emerged in the review of literature for this study. To supplement this and, importantly, for the chart to be a more accurate reflection of the skills requirements for the digital era academic library environment, the ACRL *2012 top ten trends in academic libraries* (ACRL Research Planning & Review Committee, 2012) was consulted and the following three items of disciplinary knowledge were added: scholarly communication, e-resources collection development, and research support librarianship. Further, where necessary, knowledge and skills included in the categories were "iteratively extended and refined" (Wise et al., 2011, p. 275) during the analysis of the data — that is, as the researcher worked through the contents of the job requirements sections of the advertisements and the transcripts of the interviews, where necessary, required skills sets were grouped together under a general attribute if they were part of a generic group of skills or separated into specific attributes if they fell into distinct skills requirements, depending on the pattern that unfolded as the researcher meticulously worked through the content of the job requirements data collected.

Table 2. Knowledge and skills requirements in the digital era academic library in South Africa.

Disciplinary knowledge	Job adverts		Interviews		Generic skills	Job adverts		Interviews		Personal competencies	Job adverts		Interviews	
	Freq.	%	Freq.	%		Freq.	%	Freq.	%		Freq.	%	Freq.	%
Traditional LIS resources, services and functions	71	33%	21	46%	General managerial/supervisory skills	48	19%	6	11%	Enthusiasm and initiative	18	26%	2	9%
Experience	42	19%	3	6.5%	Communication (written and oral)	33	13%	6	11%	Leadership	13	19%	3	13.6%
Technology associated with LIS applications in the digital era	40	18%	3	6.5%	Interpersonal skills	26	10.2%	2	3.5%	Capacity to work independently	11	16%	--	--
Scholarly communication	18	8%	3	6.5%	General computer skills and computer literacy (including social media skills)	25	10%	9	16%	Ability to work under pressure	11	16%	1	4.5%
e-Resources collection development	13	6%	5	11%	Client service orientation	23	9%	4	7%	Ability to respond to others' needs	6	8.7%	3	13.6%
Research support librarianship	11	5%	2	4%	Teamwork	21	8%	3	5%	Adaptability and flexibility	5	7.2%	3	13.6%
Subject knowledge with professional LIS skills	10	4.6%	1	2%	Critical and analytical thinking	15	6%	--	--	Capacity for continuous learning	2	3%	6	27%
Digitization (creating digital content)	6	3%	3	6.5%	Teaching and coaching	12	5%	9	19%	Fostering change	1	1.5%	1	4.5%
Curation of digitized content (including metadata creation and management)	4	2%	3	6.5%	Budgeting	8	3.1%	1	1.5%	Reflective thinking	1	1.5%	1	4.5%
Research data services (including collection, metadata creation and preservation for future use)	2	1%	2	4%	Problem solving	7	2.7%	1	1.5%	Self-motivation	1	1.5%	2	9%
					Project management	6	2.3%	1	1.5%					
					Accuracy and attention to detail	5	2%	2	3.5%					
					Strategic planning	5	2%	2	3.5%					
					Branding and marketing	4	1.5%	2	3.5%					
					Information literacy	4	1.5%	3	5%					
					Research skills	3	1.2%	2	3.5%					
					Writing/editing skills	2	1%	2	3.5%					
					Experience	2	1%	--						
					Conflict management skills	2	1%	--						
					Continuing professional development	2	1%	--						
					Performance management and evaluation skills	1	0.5%	1	1.5%					
Total	217	100%	46	100%		254	100%	56	100%		69	100%	22	100%

The advertisements were scrutinized for requirements and these were allocated to the skills categories on the chart. Data was gathered from the requirements sought by the employers and not from the duties or responsibilities of the post, unless this referred to a particular requirement (Orme, 2008, p. 624). Each requirement was recorded on the tally chart. Once this was done for all the advertisements, the counts were totalled to produce frequency counts allowing for "an assessment of the relative importance of the different requirements and the areas into which they fall" (Orme, 2008, p. 624). A total of 71 advertisements were analysed. A similar process was applied to the data collected from the six interviews (a total of about 6 h of interviewing) except here the researcher had to wade through much qualitative data and had to be mindful of not capturing repetitions. The frequency counts (together with percentages) from both the data sets are reflected inTable 2. Fig. 1, using a bar graph, captures the totals (and corresponding percentages) for each of the skills categories. Juxtaposing the totals from the two data sources (job advertisements and interviews) on the graph allows for comparison of trends between the two data sets, for purposes of triangulation.

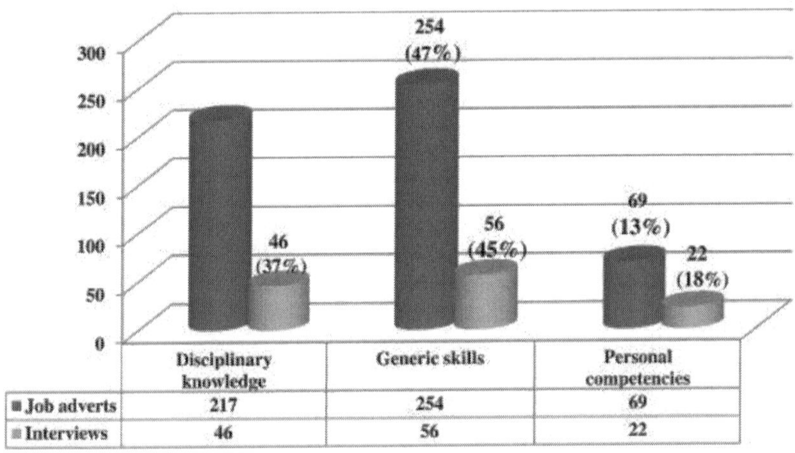

Figure 1. Knowledge and skills requirements in the digital era academic library in South Africa.

DISCUSSION, CONCLUSIONS AND LESSONS FOR THE WIDER ACADEMIC LIBRARY SKILLS STUDY

This section discusses the findings from this exploratory study in relation to the literature reviewed and based on this discussion draws conclusions in response to the research question: What key knowledge and skills are required for LIS professionals to effectively and efficiently practise in a digital era academic library in South Africa? A secondary purpose of this preliminary investigation was to tease out some of the parameters for the wider study targeting the development of a comprehensive skills statement for higher education libraries in South Africa. Hence this section also looks at some of the lessons gained from this exploratory investigation for the wider study.

DISCUSSION AND CONCLUSIONS

As revealed in the literature (Nonthacumjane, 2011, Orme, 2008, Partridge and Hallam, 2004 and Wise et al., 2011) and reflected in Table 1, in this study too it emerges that a variety of skills and competencies are required of the modern LIS professional. A key convergence in findings between this South African study and similar ones in the USA, United Kingdom and Australia cited in the review of literature, is that generic skills emerged as being the most required skills set (see Fig. 1). Table 1, which captures key trends from the literature relating to academic library knowledge and skills requirements, shows a 55% preponderance of the trend of employers increasingly emphasizing generic skills as priority skills requirements. At the same time, Orme (2008, p. 626) observes that despite the 'move to the generic' argument in the literature (Kennan et al., 2006 and Wise et al., 2011), and also evident in the findings of the study reported here, disciplinary knowledge is still valued by LIS employers, as evidenced in their US study by a strong presence of disciplinary skills in the top twenty list of frequently sought skills (see alsoTable 1 — 45% preponderance of this trend reflected in the literature). The current study lends weight to this argument when one observes in Table 2 the high frequency counts against, for example, 'traditional LIS resources, services and functions' (71 and 21, respectively) in both the job advertisement and interview data sets. This South African study, like those abroad, also saw personal skills, while sought by employers, lagging behind disciplinary knowledge and generic skills in popularity (see Fig. 1). Findings from the literature, summarised in Table 1, reiterate this point by reflecting this as a notable trend in the literature.

'Technology associated with LIS applications in the digital era' as a required knowledge/skills set achieved a high frequency count in the job advertisement data set (see Table 2). This is not surprising given the impact of technology on the academic library of the digital era, as detailed in the literature (Choi and Rasmussen, 2009, Luce, 2008, McCarthy, 2005, O'Connor and Au, 2008, Ocholla and Shongwe, 2013 and Reeves and Hahn, 2010) and reflected in a 64% preponderance of this trend in the literature (see Table 1). While this exploratory study conflated all technology associated with LIS applications into one skills category, in the wider study envisaged it would be useful to separate this into knowledge of and familiarity with systems software on the one hand, and technical skills on the other. Such technical skills would include digital library architecture and software, technical and quality standards, HTML coding, Web mark-up languages such as SGML and XML, and possibly even some skills in programming and scripting languages. Such a separation would be useful to determine to what extent IT technology skills are becoming a necessity in the education and training of academic library LIS professionals. This IT skills set becomes particularly relevant when one considers that the literature is very emphatic about the fact that the digitally oriented academic library is a renewed conceptualisation of traditional LIS resources and services that is now underpinned and driven by new technologies (Choi and Rasmussen, 2009, Gerolimos and Konsta, 2008, McCarthy, 2005, Middleton, 2003 and Missingham, 2006) or that "LIS [particularly in the digital academic library environment] is becoming an increasingly technology-driven profession" (Riley-Huff & Rholes, 2011, p. 129). This would also explain 'Technology associated with LIS applications in the digital era' following closely on the heels of 'Traditional LIS resources, services and functions' in the upper end of frequency counts in the job advertisement data set in Table 2.

The high job advertisement frequency count against 'Experience" in the disciplinary knowledge skill set (42 — see Table 2) implies, as Orme (2008, 629) suggests, that for employers knowledge, skills and personal attributes alone are "insufficient without an ability to demonstrate their practical application". The job advertisement frequency counts for, 'Scholarly communication', 'e-Resources collection development' and 'Research support librarianship' (see Table 2) indicate that these skills areas have taken hold in the modern academic library in South Africa and are on an upward trajectory as suggested in the literature (Choi and Rasmussen, 2009, Luce, 2008 and Nonthacumjane, 2011) about academic libraries in other parts of the world. 'Scholarly communications', particularly, received quite a boost in frequency counts from among 2013 advertisements, indicating an increasing demand in the digital academic library environment for knowledge of new trends in scholarly communications. Combining subject knowledge with professional LIS skills is somewhat of a grey area with just one interviewee in the current study claiming that this is "the ideal" which the LIS service should strive for. Partridge and Hallam (2004) report that "'subject discipline' may becoming more and more important" ...especially in fields such as law and medicine where LIS professionals need to "develop or acquire skills and knowledge unique to that field". Not co-incidentally, the frequency count of 10 for subject knowledge requirement inTable 2 emanates largely from law libraries. The rest of the interviewees (five) believed that while having the 'subject discipline' was an advantage, this is "hardly a reality—it is difficult to find a medical graduate working in a medical library".

The emergence on the radar screen of new skills types such as 'Curation' and 'Research data services', albeit with low frequency counts (see Table 2), represent in the South African context too the "emerging trend of research data management and curation" (Wise et al., 2011, p. 279) accompanying the emergence of eScience and eResearch (Luce, 2008), as has been observed by researchers in academic LIS services internationally. South African academic libraries too, like their overseas counterparts are being confronted, in a context of a shortage of skills and training in this area, with the challenge of developing data-archiving infrastructure for the description, management, access and sharing of data (University of Melbourne, 2008). The researcher is convinced that in another year or so, the frequency counts in these new skills areas (digitization, curation, research data service) would rise significantly.

While the generic skills set in Table 2 is led by 'General managerial/supervisory skills' with a frequency count of 48 (in the job advertisement data set), this high figure might be the result of the fact that this skills category, like the 'Traditional LIS resources, services and functions' category in the disciplinary knowledge section, was a generic skills category that pooled together a number of different skills. In the wider study planned, these aggregated categories should perhaps be thinned out for a more accurate reflection of knowledge and skills trends. Notwithstanding this, the generic skill 'Communication' which traditionally is reflected in the literature as the most highly ranked generic skill (Gerolimos and Konsta, 2008, Middleton, 2003 and Orme, 2008), does reflect in Table 2 as a noticeably highly ranked generic skill, both in the job advertisement as well as in the interview data sets. Mention needs to be made of 'Interpersonal skills', 'General computer skills and computer literacy', 'Client service orientation' and 'Teamwork' — these are generic skills which also notched up significant frequency counts (see Table 2) lending weight to the 'move to the generic' trend evident in the literature (see Table 1) as well in the findings reported in this paper (see Fig. 1 andTable 2). This increasing tendency towards a demand for skills relating to generic capabilities, over and above disciplinary or professional knowledge and skills, is also reflected by the longer list of skills requirements in the generic skills set compared to the other two (see Table 2). The 2013 job advertisements, particularly, surfaced noticeable requests for generic skills relating to project management, strategic planning, branding and marketing, budgeting, performance management and evaluation, etc. (see Table 2). Digital academic libraries are expensive enterprises, and hence it is not surprising that these generic skills are in demand to ensure the efficacy of these enterprises.

It would seem that the knowledge and skills required for LIS professionals to effectively and efficiently practise in a digital era academic library in South Africa are a blend of discipline-specific knowledge, generic skills and personal competences, with some of these skills types being more valued in this context than others. At the same time, these LIS professionals, like their counterparts in other parts of the world, are also being challenged by new and emerging skills requirements. Many of the latter are a re-conceptualisation of traditional LIS skills using new technologies and hence it is important for the new generation LIS professionals to use this knowledge base to adapt existing skills to respond to new 'problems' in a working environment that is constantly changing. LIS education and training in South Africa too needs to evolve to meet the challenges of the new knowledge and skills requirements of the digital age academic library in South Africa which is a major employer of LIS graduates.

While the intention of this preliminary study was to ascertain an initial picture of key knowledge and skills sets required for LIS professionals in this environment, a secondary intention was to tease out of this exploratory investigation some of the parameters for the wider study targeting the development of a comprehensive skills statement for higher education libraries in South Africa. Hence this paper ends by highlighting lessons for the wider study.

Lessons for the wider study

This exploratory study was useful in revealing the limitations of relying on a single data source for job advertisements, no matter how good a source it is. It was also instructive in revealing that in the current digital age, online sources of job advertisements are a must if one wants a "comprehensive picture of the job market" (Reeves & Hahn, 2010, p. 105). Hence the envisaged wider study would need to target multiple sources of job advertisements, such as other newspaper titles, online mailing lists (particularly those of the professional body (LIASA) and relevant websites. This is likely to provide a more complete picture of the LIS employment market in South Africa.

While this small study depended on manually tallying the frequency of skills requirements identified through content analysis, the bigger study (with bigger data sets) could make use of content analysis software such as *Provalis Simstat* and *Wordstat* (Wise et al., 2011, p. 275) used in other successful job advertisement studies. The interviews too were time-consuming, geographically restricting (to the Western Cape) and its more qualitative nature did not blend very naturally with the type of data collected from the job advertisements which lent itself more to quantification and hence more convenient analysis during the triangulation process. Moreover, the interviews, due to limitations to the number that can be conducted over a specified period of time, did not prove, in the end, to be such a productive source of research data for this type of study. Based on this research experience and the knowledge gained from this exploratory exercise, the researcher would suggest that the wider study, instead of gathering data via semi-structured interviews, make use of quantitative survey data collected via the use a structured Web-based questionnaire administered to LIS professional staff from academic libraries scattered across the country, for a truly national picture and a more efficient triangulation of data sources. Content analysis software could be used to analyse the much larger volume of data that is intended for collection. The use of data analysis software with a larger volume of data would also allow for more than just frequency counts. Other more sophisticated analyses such as co-occurrence and multivariate analyses (Wise et al., 2011, p. 283) could be employed, thus allowing for deeper analysis and a more complete picture of the requirements of employers, employment opportunities and emerging trends in the LIS employment market.

REFERENCES

1. ACRL Research Planning and Review Committee (2012). 2012 top ten trends in academic libraries: A review of the trends and issues affecting academic libraries in higher education.

2. Choi, Y., & Rasmussen, E. (2006). What is needed to educate future digital librarians: A study of current practice and staffing patterns in academic and research libraries. D-Lib Magazine, 12(9) (Retrieved May 14, 2012, from http://www.dlib.org/dlib/ september06/choi/09choi.html).

3. Choi, Y., & Rasmussen, E. (2009). What qualifications and skills are important for digitallibrarian positions in academic libraries? A job advertisement analysis. The Journal of Academic Librarianship, 35(5), 457–467.

4. Gerolimos, M., & Konsta, R. (2008). Librarians' skills and qualifications in a modern information environment. Library Management, 29(8/9), 691–699.

5. Kennan, M.A., Cole, F., Willard, P., Wilson, C., & Marion, L. (2006). Changing workplace demands: What job ads tell us. Aslib Proceedings: New Information Perspective, 58(3), 179–196.

6. Luce, R. E. (2008). A newvalue equation challenge: The emergence of eResearch and roles for research libraries. Retrieved January 22, 2013, from. http://www.clir.org/pubs/ reports/pub142/luce.html

7. McCarthy, J. (2005). Planning a future workforce: An Australian perspective. New Review of Academic Librarianship, 11(1), 41–56.

8. Middleton, M. (2003). Skills expectations of library graduates. New Library World, 104(1184/1185), 42–56.

9. Missingham, R. (2006). Library and information science: Skills for twenty-first century professionals. Library Management, 27(4/5), 257–268.

10. Neuman, W. L. (2006). Social research methods: Qualitative and quantitative approaches. Boston: Pearson.

11. Nonthacumjane, P. (2011). Key skills and competencies of a newgeneration of LIS professionals. IFLA Journal, 37(4), 280–288.

12. O'Connor, S., & Au, L. -C. (2008). Steering a future through scenarios: Into the academic library of the future. The Journal of Academic Librarianship, 35(1), 57–64.

13. Ocholla, D., & Shongwe, M. (2013). An analysis of the library and information science (LIS) job market in South Africa. South African Journal of Libraries and Information Science, 79(1), 35–43,

14. Orme, V. (2008). You will be ...: A study of job advertisements to determine employers' requirements for LIS professionals in the UK in 2007. Library Review, 57(8), 619–633.

15. Partridge, H., & Hallam, G. (2004). The double helix: A personal account of the discovery of the structure of [the information professional's] DNA. Paper presented at the ALIA 2004 Biennial Conference, Gold Coast, Australia, 21–24 September 2004 Retrieved January 25, 2013, from http://conferences.alia.org.au/alia2004/pdfs/ partridge.h.paper.pdf.

16. Reeves, R. K., & Hahn, T. B. (2010). Job advertisements for recent graduates: Advising, curriculum, and job-seeking implications. Journal of Education for Library andm Information Science, 51(2), 103–119.

17. Riley-Huff, D. A., & Rholes, J. M. (2011). Librarians and technology skill acquisition: Issues and perspectives. Information Technology in Libraries, 30(3), 129–140.

18. Tammaro, A.M. (2007). A curriculum for digital librarians: A reflection on the European debate. New Library World, 108, 229–246.

19. University of Melbourne (2008). Future skills for academic librarians: research and research and training paper for Information Futures Commission. Retrieved March 15, 2012,

20. Welman, C., Kruger, F., &Mitchell, B. (2005). Research methodology (3rd ed.). Cape Town: Oxford University Press

21. Wise, S., Henninger, M., & Kennan, M.A. (2011). Changing trends in LIS job advertisements. Australian Academic and Research Libraries, 42(4), 268–295.

11

CHAPTER

Factors that Increase the Probability of a Successful Academic Library Job Search

Max Eckard[a], , Ashley Rosener[b, 1], , Lindy Scripps-Hoekstra[c, 2],

[a] *Metadata & Digital Curation Librarian, University Libraries, Grand Valley State University, Allendale, MI 49401, USA*

[b] *Liaison Librarian in Professional Programs, University Libraries, Grand Valley State University, Grand Rapids, MI 49504, USA*

[c] *Liaison Librarian in Liberal Arts Programs, University Libraries, Grand Valley State University, Allendale, MI 49401, USA*

ABSTRACT

Finding a position in an academic library can be challenging for recent Library and Information Science (LIS) graduates. While LIS students are often encouraged to seek out experience, network, and improve upon their technology skills in hopes of better improving their odds in the job market, little research exists to support this anecdotal advice. This study quantifies the academic and work experiences of recent LIS graduates in order to provide a better understanding of what factors most significantly influence the outcome of their academic library job searches. The survey results demonstrate that the job outlook is most positive for candidates who applied early, obtained academic library experience (preferably employment), participated in professional conferences, and gained familiarity with committee work.

Keywords: Job hunting; Academic libraries; Education

INTRODUCTION

Finding a position in an academic library can be challenging for the newly minted library and information science (LIS) graduate. Fortunately, there is no shortage of job-seeking advice for prospective candidates. From librarian blogs to professional magazines and websites such as "I Need a Library Job" (www.inalj.com), suggestions and recommendations abound. In their 2010 C&RL News piece, Cannady and Newton recommended making the "best of the worst of times" (p. 210) and gave specific advice for each phase of the job search: application, phone interview, and face-to-face interview.Singleton (2003) recommended building a network of professional contacts and being prepared to show effectiveness in professional assignment, professional development, and service. In "Academic Library Job Search Blues," Baker (2010) advised making oneself more marketable by finding internships, becoming involved in organizations, and taking a balanced blend of classes.

While this advice may be sound, it is largely based upon the subjective experiences of the authors. The overabundance of such anecdotal advice can make a difficult job search even more challenging. With the variety of choices in coursework, internships, and other opportunities students are advised to pursue during their short time in graduate school, how will they know which activities are the most beneficial for ensuring a successful academic library job search? The objective of this study is to quantify the experiences of recent LIS graduates to better understand what factors might help job seekers obtain their first professional position. By comparing the academic library job search outcomes of survey respondents, this study shows that applying early, obtaining academic library experience (preferably employment), participating in professional conferences, and gaining familiarity with committee work increases the probability of being able to obtain a first post-degree position. The results of this study also identify trends in LIS students' graduate school involvement (academic and extracurricular) and recent LIS graduates' perspectives of the academic library job market and the process of securing a job.

LITERATURE REVIEW

The labor market for LIS graduates has been described, in general, as "relatively 'recession proof'" (Morgan & Morgan, 2009, p. 299), and indeed, librarianship has endured the Great Recession of the late 2000s. The 2012–13 edition of the Occupational Outlook Handbook stated that employment of librarians is actually expected to grow by 7% from 2010 to 2020 (U.S. Department of Labor & Bureau of Labor Statistics, 2012, 2013). While slower than average for all occupations, this is growth nonetheless. According to an analysis done in 2012, salaries are up 5% while unemployment for recent graduates held at 6%, suggesting that positions are available, with similar numbers for 2013 (Maatta, 2012 and Maatta, 2013).

Academic Library Job Market

According to the Bureau of Labor Statistics, later in the decade prospects should be even better as older library workers retire and population growth generates openings (U.S. Department of Labor & Bureau of Labor Statistics, 2012, 2013). However, it should be noted that this optimistic prediction is reminiscent of the early 2000s expectation that the availability of academic library jobs would increase "due in large part to the 'graying' of the profession" (p. 408), an estimation that never came to fruition because budget constraints, especially in local government and educational services, slowed demand for librarians (Tewell, 2012).

The academic library job market for recent LIS graduates is competitive, and for those lacking significant practical experience, it is a "potentially insurmountable challenge" (p. 422) (Tewell, 2012). Nearly three-quarters of academic librarian positions in 2011 preferred or required work experience (Triumph & Beile, 2011). Also, in 2012 almost three-quarters of academic library jobs were non-entry level, and of those 26% of all job advertisements were administrative, creating a tough market for recent graduates (Tewell, 2012).

Search Committee Perspectives

The academic library job search and hiring process provides an additional challenge for recent graduates. During times of economic distress when employment opportunities are few and the number of applicants increases, candidates must be aware of the rigorous academic library job search process (Durán, Garcia, & Houdyshell, 2009). In their survey of search committees, Hodge and Spoor found that 78% of respondents are receiving more applications per position opening now than in previous years, but 80% do not interview more candidates (2012). Search committee members can suffer from fatigue while reviewing such large numbers of applications and may overlook qualified candidates (Howze, 2008).

The Association of College and Research Libraries' Discussion Group of Personnel Officers "agreed that previous library experience was an important requisite for an entry level position. The majority also indicated that the experience should be in an academic library" (Neely, 2011, p. 4). Previous work experience is an indicator of future job performance, as Wheeler, Johnson and Manion highlighted when they suggested that questions about a candidate's experience are more effective than situational questions(2008). Demonstrated performance of job requirements is very important. In their 2010 study, Wang and Guarria found that 90% of the 243 survey respondents (individuals who served on faculty search committees) believed that a demonstrated ability to perform job requirements was very to extremely important (p. 83).

In addition to work experience and the ability to perform job requirements, candidates must demonstrate they are leaders, not just workers. Search committees want librarians who are creative, proactive, risk takers, innovators, independent yet collaborative, lifelong learners, and visionaries (Harralson, 2001). Reeves and Hahn (2010) reminded LIS students that employers prefer individuals who have good communication skills, work well with others, take initiative, are adaptable and dependable, and have a "service orientation, a predilection for collaboration and cooperation [and] a penchant for participating in teams" (p. 118). Employers also look for applicants who can acclimate quickly to organizational culture. In Wang and Guarria's survey, over 90% of the 243 survey respondents said potential fit is very or extremely important in an academic library(2010). For recent graduates with less work experience, this further emphasizes the importance of potential fit with the organizational culture.

METHODOLOGY

In order to learn more about the graduate school experiences and job search successes of recent LIS graduates, an electronic survey was created using SurveyMonkey (Appendix A). In March 2013, a link to the survey was emailed to 2008–2012 graduates from the LIS programs at the University of Illinois at Urbana-Champaign and North Carolina Central University. The survey link was also emailed to members of the ALA New Members' Round Table (NMRT) listserv, distributed on index cards to ACRL 2013 conference attendees during a related poster presentation, and electronically posted on the ACRL New Member Discussion board. At the request of the LIS program at Dominican University, the survey link was not emailed to their graduates until July 2013. In addition to being the three universities from which the three researchers graduated, these institutions represent a cross section of different LIS programs around the United States in that they have different U.S. News & World Report rankings, program formats, enrollment numbers, and local job markets. Respondents from the three universities and the NMRT listserv provided a diverse sample of recent LIS graduates from ALA-accredited programs around North America. While the respondents graduated in different years throughout the last decade, the survey questions only addressed their time in graduate school.

The survey questions were divided into seven primary categories: basic information, job search, professional effectiveness, professional development, service, technological competency, and previous careers. Questions focused on the student's graduate program, the parameters of their job search, their academic and work experience, as well as other skills or professional involvement that might impact their success in landing a first job. In addition to these questions, the survey concluded with an open-ended response question that asked, "What advice do you have for current LIS students?" This question and all other free-response questions were coded by the researchers and differences in coding were resolved by mutual agreement. The inter-coder reliability averaged Kappa = .9, which represents an almost perfect agreement according to Landis and Koch (1977, p. 165).

Limitations

The results are based upon a limited sample size of recent LIS graduates since the survey was sent to three specific universities. This limitation was offset to a certain extent by also sending the survey to the NMRT listserv, posting it on the ACRL New Member Discussion board, and passing out links at the 2013 ACRL Conference, which allowed for gathering responses from graduates from many other LIS schools. Due to the numerous avenues by which the survey link was dispersed, it is not possible to calculate the response rate. The results are also limited due to the fact that survey questions were optional and not all elicited an adequate number of responses for the following analysis.

RESULTS

Out of 360 total survey respondents, 56% (N = 201) indicated that they wanted to work in academic libraries. Eighty-two respondents wanted to work in public libraries with smaller numbers selecting "special libraries" or "other". The survey respondents provide a varied sample of LIS graduates. The 201 respondents who selected academic librarianship were recent LIS graduates from the years 2005–2013 with the highest percentage (28%) graduating in 2011. Respondents represented 33 different LIS programs with the highest number of students graduating from the University of Illinois (56) and Dominican University (39).

SUCCESSFUL RESPONDENTS

Successful respondents are defined as the subset of survey respondents who were able to find a full-time or part-time, tenure-track or professional academic library job after graduation. Respondents were asked about the outcome of their job search and 186 (n = 186) individuals responded to this question. Sixty-eight percent (126) indicated that their job search was successful and 23% (60) reported an unsuccessful job search. In the following analysis, percentages are based upon the number of successful respondents who answered each question. As this study's aim is to determine the factors that lead to success in obtaining a job, our analysis focuses only on the successful respondents.

Basic information

Over 60% (75) of our successful respondents were full-time students. In terms of program format, 44% (56) attended their program in-person, 30% (43) enrolled in a hybrid program, and 24% (30) completed their degree online.

Job search

Respondents were asked when they began applying for jobs, and 62% (75) reported having started the application process four months or more prior to graduation. In terms of the number of positions applied for, the highest percentage (32%) of successful respondents applied for five or fewer jobs, while the lowest percentage (8%) of respondents reported applying to over 50 positions (Fig. 1). Almost half of successful respondents (52) found a position before graduating, and 40 respondents found a position within one to three months after graduation.

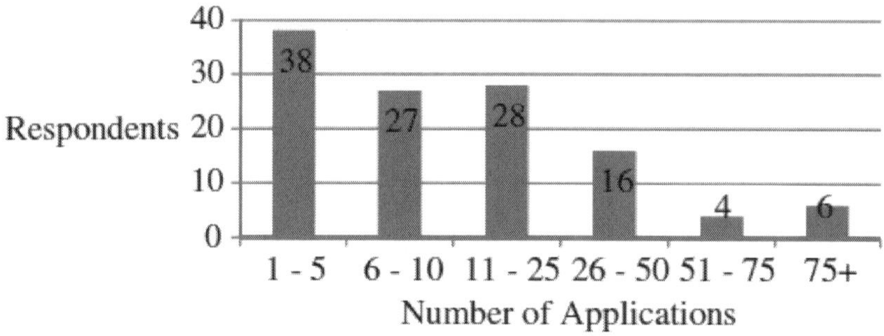

Figure 1. Number of successful applications.

Just over 50% of respondents (65) were limited to a specific geographic area with 56 reporting the ability to move for a position. Almost half found a job as a reference and instruction librarian (Fig. 2). The next most common type of position was one in technical services.

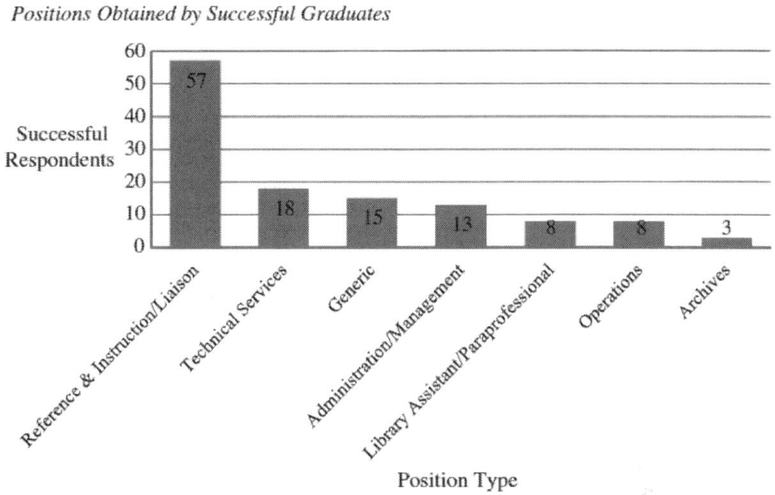

Figure. 2. Positions obtained by successful graduates.

The types of positions reported varied in status, and 86% (108) reported landing full-time positions. Over half of successful respondents (62) described their position as "professional," and a third (42) described their position as "faculty" or "tenure-track".

Professional effectiveness

Survey respondents were also asked to select their coursework choices and extracurricular activities from a provided list (respondents could select multiple options). Results indicated that 82% (103) took at least one class in academic librarianship, 66% (83) worked in an academic library, and approximately 51% (64) had an internship in an academic library. Some respondents also gained experience by volunteering at libraries. Of these respondents, 23% (29) volunteered in a public library, 15% (19) in a special library, and 8% (10) in an academic library.

Professional development

Approximately 67% of respondents reported having attended conferences (84) during graduate school. A similar percentage attended workshops and seminars (77). Almost 25% (30) participated in conferences. Finally, 14% (18) had either authored or co-authored a publication of some type.

Service

Over 50% (81) of the respondents were involved in library associations and/or were active in library student groups (65).

Technological competency

In response to a general question about participation in elective technology courses while in graduate school and self-reported technological competency level, seventy-six percent (96) of respondents rated themselves as either "Competent" or "Very competent" with technology.

Previous careers

Just over 50% (64) came to librarianship as a second career and 86% (75) of them worked for ten years or less in their previous careers. Of these respondents, 66% (58) felt their previous job helped them land their library job. Respondents were asked to name their previous career(s), and their answers were coded based upon the occupational groups identified by the Bureau of Labor Statistics' in the Occupational Outlook Handbook. For successful respondents, the most common previous careers were in the fields of education, training, or libraries.

Respondent advice

The survey concluded with an open-ended response question crafted to elicit perceptions on how to be successful in securing a job. Respondents were asked what advice they had for current LIS graduate students (Fig. 3). Answers varied, but some pieces of advice came up repeatedly. Of successful respondents, 40% felt that experience was important and 18% recommended networking. Twelve percent recommended job market preparation, perhaps in the form of help with formatting and writing cover letters and resumes. Just 11% pushed for graduates to increase their technological savvy, and only four percent respondents recommended that students be willing to move for a job.

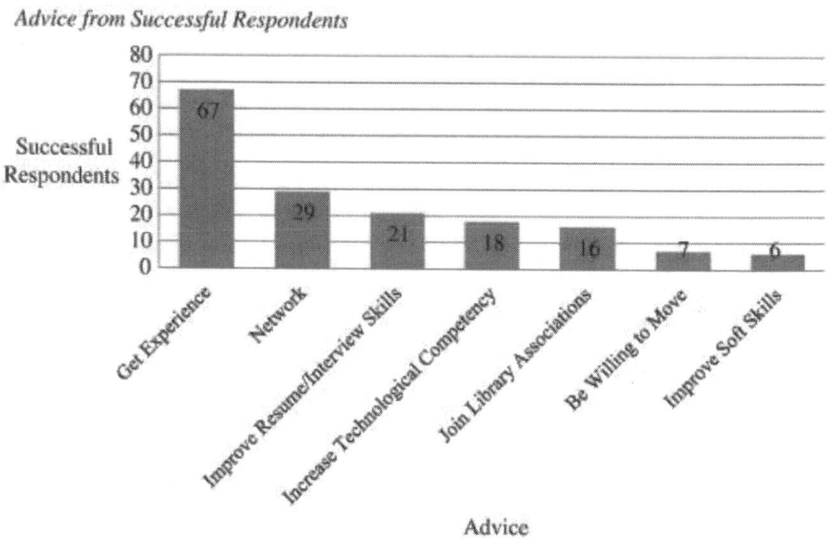

Figure. 3. Advice from successful respondents.

DISCUSSION

We compared successful and unsuccessful job seekers in order to determine whether any trends existed for either group. Overall, the two groups were largely similar. Of the seven question categories, only certain factors in job search, professional effectiveness, professional development, and service made a significant difference in improving a candidate's odds of success in landing a job. Among these factors were academic library experience, committee work, conference attendance, and publications. These factors significantly increased or decreased the odds of getting a job after graduation; odds ratios were calculated and analyzed for respondents using 95% confidence intervals (Ekstrom & Sorenson, 2011, p. 321).

Job Search

Recent graduates were asked when they began applying for jobs. The odds ratio (6.78) showed that candidates who began applying for jobs four to six months before graduation were nearly seven times ($p = .0092$) more likely to obtain a job than candidates who did not. Based on the confidence intervals (1.26, 5.6), we can be 95% confident that the entire population of LIS graduates who begin applying for jobs four to six months before graduation are between 1.26 and 5.6 times more likely to get a job compared to those who do not. Applying for jobs four to six months before graduation increased a candidate's odds of success more than any other factor.

Professional Effectiveness

Having any academic library experience increased the odds that a job seeker would be successful, although not all types of experience were significant. The odds ratio of getting a job for candidates who had been employed in academic libraries compared to those who had not was 4.71 ($p < .0001$). The 95% confidence interval for this odds ratio is between 2.41 and 9.22. Participation in an academic library internship or practicum also significantly improved the odds that a recent graduate would successfully find a job. Candidates who had participated in an internship or practicum improved their odds of success by 2.75 ($p = .0027$) compared to candidates who had not completed an internship or practicum, with the 95% confidence limits for the entire population of recent graduates ranging from 1.41 to 5.38.

Professional Development

Recent graduates were also asked about professional development experiences. Candidates who had attended conferences increased their odds of success by 3.33 ($p = .0002$) when compared to candidates without this experience, with the 95% confidence limits for the entire population being between 1.76 and 6.33. Candidates who had attended workshops and seminars increased their odds by 2.05 ($p = .023$), with 95% confidence limits from 1.1 to 3.83. Finally, candidates who had publications increased their odds by 4.83 ($p = .0246$), with 95% confidence limits from 1.08 to 21.6.

Service

The last category that significantly impacted success was service. Recent graduates were asked which service opportunities they participated in while in graduate school including association membership, student groups, committee work, volunteering, and fundraising. Of these, only committee work was significant. Candidates who did committee work increased their odds by 3 ($p = .0274$), with 95% confidence limits between 1.09 and 8.23.

CONCLUSION

This study sought to provide a better understanding of which quantifiable factors can significantly influence an academic library job search. Based on these results, time of application, academic library experience, committee work, conference attendance, and authoring or co-authoring a publication appear to increase the odds of a recent graduate's ability to break into the field.

In addition to the factors that improve the odds of success in a job search, this study also found other factors addressed in the survey did not appear to significantly affect a graduate's chance of securing a job. Among these factors were: program format, graduation date, grade point average, enrollment status, study abroad experience, independent study, and technological skills. It is difficult to quantify whether potential fit impacts success in landing a library position, but the literature suggests that a candidate's personality and potential fit with a library is a large consideration when hiring committees make their hiring decision. Due to the fact that successful and unsuccessful respondents were fairly similar, it could be deduced that potential fit plays an important role in the hiring process.

While students are ultimately responsible for preparing themselves for a job search, LIS programs can do more to assist them. Based on this research, we recommend LIS programs provide opportunities for students to obtain some form of academic library work experience. Considering the importance of early application, LIS advisors should also ensure that students are prepared for the unique considerations of the academic library job search process. LIS programs should also focus on connecting students to professional development and publication opportunities.

Our research represents an initial look into several factors in the academic library job search, yet many other variables warrant further exploration. More research could be conducted on what job market preparations are offered at different LIS programs and the potential impact of online portfolios and professional social networking. While this study focused on academic library job seekers, public and other types of libraries have yet to be examined. Further research could also explore the experiences of unsuccessful library school graduates in order to determine whether trends exist.

For recent LIS graduates hoping to obtain academic library positions, the outlook appears positive for candidates who begin applying for jobs four to six months before graduation, and who are able to procure academic library experience (preferably employment), become involved with professional conferences, and gain familiarity with committee work. While these conclusions may not surprise many in the field, this study represents a quantitative understanding of oft-given job seeking advice. The advice given by 40% of successful respondents to get experience was sound; as one respondent stated, "Experience makes all the difference!"

ACKNOWLEDGMENTS

Special thanks to the Grand Valley State University Statistical Consulting Center.

appendix a. survey instrument

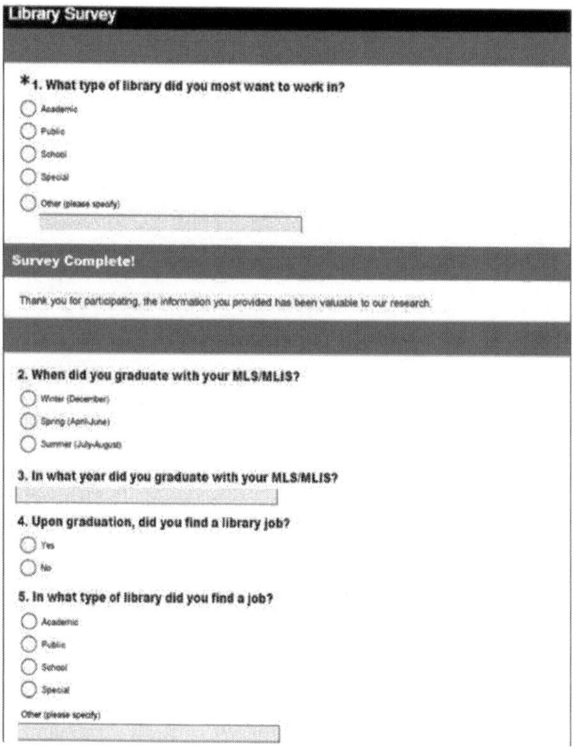

Library Survey

*1. What type of library did you most want to work in?

- Academic
- Public
- School
- Special
- Other (please specify)

Survey Complete!

Thank you for participating, the information you provided has been valuable to our research.

2. When did you graduate with your MLS/MLIS?

- Winter (December)
- Spring (April-June)
- Summer (July-August)

3. In what year did you graduate with your MLS/MLIS?

4. Upon graduation, did you find a library job?

- Yes
- No

5. In what type of library did you find a job?

- Academic
- Public
- School
- Special

Other (please specify)

6. Is the job full time or part time?

- Full Time
- Part Time

7. What is the classification of your position?

- Professional
- Faculty/Tenure Track
- Para-professional

Other (please specify)

Job Search

8. When did you begin applying for jobs?

- More than 7 Months Before Graduation
- 4-6 Months Before Graduation
- 2-3 Months Before Graduation
- 1 Month Before Graduation
- Less than 1 Month Before Graduation
- After Graduation
- Other (please specify)

9. How many library positions did you apply for?

- 1-5
- 6-10
- 11-25
- 26-50
- 51-75
- Over 75
- Other (please specify)

10. What percentage of job applications would you ESTIMATE resulted in advances to the next level in the hiring process (phone/in-person interviews, etc.)?

11. How long after graduation did it take you to find a library job?

- Before graduation
- 1-3 Months
- 4-6 Months
- 7-9 Months
- 10-12 Months
- More than a year

12. What was the title of your first library job after graduation?

13. Were you limited to a specific geographic area when applying for library jobs?

- Yes
- No

14. Which school did you graduate from with your MLS/MLIS degree?

- University of Illinois Champaign-Urbana
- Dominican University
- North Carolina Central University
- Other (please specify)

15. Which type of degree did you obtain?

- ◯ MLS
- ◯ MLIS
- ◯ MIS
- ◯ Other (please specify)

16. What was your graduating GPA?

17. What was your enrollment status?

- ◯ Full time student only
- ◯ Part time student only
- ◯ Some combination of full and part time student

18. What was the format of your program?

- ☐ In person
- ☐ Online
- ☐ Hybrid

19. In which of the following experiences did you participate in during graduate school (check all that apply)?

	Public Library	Academic Library	School Library	Special Library	Other	None
Coursework	☐	☐	☐	☐	☐	☐
Employment	☐	☐	☐	☐	☐	☐
Internship/Practicum	☐	☐	☐	☐	☐	☐
Volunteer	☐	☐	☐	☐	☐	☐
Study Abroad	☐	☐	☐	☐	☐	☐

Other (please specify)

20. Please indicate whether you took multiple elective courses in technology while in graduate school (yes or no), and rate your technological competency level:

	Not very competent	Somewhat competent	Competent	Very Competent
Yes	◯	◯	◯	◯
No	◯	◯	◯	◯

21. Which of the following professional development activities did you participate in while in graduate school? (check all that apply)

- ☐ Conference Attendance
- ☐ Workshops and seminars
- ☐ Independent study
- ☐ Publication
- ☐ Conference Participation (papers, presentations and poster sessions)
- ☐ Grant Writing
- ☐ Additional degrees

Other (please specify)

22. Which of the following service opportunities did you participate in while in graduate school? (check all that apply)

- ☐ Join associations
- ☐ Student groups
- ☐ Committee work
- ☐ Volunteer
- ☐ Fundraising

Other (please specify)

23. Is librarianship your first CAREER?

- ◯ Yes
- ◯ No

Previous Careers

24. Please describe your previous career(s).

25. How many years were you in your previous career(s) before you began library school?

- ○ 1-2
- ○ 3-5
- ○ 6-10
- ○ 11-20
- ○ Greater than 20

26. Do you feel that your previous career(s) helped you to land a library job?

- ○ Yes
- ○ No
- ○ Indifferent

27. Why I feel that your previous career(s) helped me to land a library job

Final Thoughts

We will be using this information to determine the extent to which graduate experiences aid in landing a job after graduation in an academic library setting.

28. Are there any other work OR life experiences/information that you typically share with potential employers at any point in the application/interview process? Please elaborate below.

29. What advice do you have for current LIS students?

30. What advice do you have for hiring managers?

Survey Complete!

REFERENCES

1. Baker, S. (2010). Academic library job search blues. Library Journal (Retrieved April 5, 2012 from http://www.libraryjournal.com/lj/communityacademiclibraries/886717-265/academic_library_job_search_blues.html.csp)

2. Cannady, R., & Newton, D. (2010). Making the best of the worst of times: Global turmoil and landing your first library job. C&RL News, 71, 205–212.

3. Durán, K., Garcia, E. P., & Houdyshell, M. L. (2009). From the inside out and the outside in: The academic library interview process in a tight economy. C&RL News, 70, 216–219 (Retrieved from http://crln.acrl.org/content/70/4/216.full.pdf+html)

4. Ekstrom, C. T., & Sorenson, H. (2011). Introduction to statistical data analysis for the life sciences. Boca Raton: CRC Press.

5. Harralson, D.M. (2001). Recruitment in academic libraries. College & Undergraduate Libraries, 8(1), 41–74.

6. Hodge, M., & Spoor, N. (2012). Congratulations! You've landed an interview. New Library World, 113(3–4), 139–161 (Retreived from http://www.emeraldinsight.com/journals.htm?articleid=17024629)

7. Howze, P. C. (2008). Search committee effectiveness in determining a finalist pool: A case study. The Journal of Academic Librarianship, 34, 340–353.

8. Landis, J. R., & Koch, G. G. (1977). Themeasurement of observer agreement for categorical data. Biometrics, 33(1), 159–174.

9. Maatta, S. L. (2012). A job by any other name: LJ's placements & salaries survey 2012.Library Journal (Retrieved from http://lj.libraryjournal.com/2012/10/placementsand-salaries/2012-survey/a-job-by-any-other-name-ljs-placements-salaries-survey- 2012/)

10. Maatta, S. L. (2013). Placements & salaries 2013: Salaries stay flat; specialties shift. Library Journal (Retrieved from http://lj.libraryjournal.com/2013/10/placements-andsalaries/ 2013-survey/salaries-stay-flat-specialties-shift/)

11. Morgan, C. H., & Morgan, J. C. (2009). The effects of entering the LIS workforce in a recession: North Carolina, 1964–2005. Library Trends, 58, 291–300.

12. Neely, T. Y. (2011). How to stay afloat in the academic library job pool. Chicago, IL: American Library Association.

13. Reeves, R. K., & Hahn, T. B. (2010). Job advertisements for recent graduates: Advising, curriculum, and job-seeking implications. Journal of Education for Library and Information Science, 51, 103–119.

14. Singleton, B. (2003). Entering academic librarianship: Tips for library school students. C&RL News, 64, 84–86.

15. Tewell, E. C. (2012). Employment opportunities for new academic librarians: Assessing the availability of entry level jobs. Libraries and the Academy, 12, 407–423 (Retrieved from http://muse.jhu.edu/journals/portal_libraries_and_the_academy/v012/12.4. tewell.html)

16. Triumph, T., & Beile, P. (2011). The trending academic library job market: An analysis of job ads from 2011, with comparisons to 1996 and 1988 studies. (Retrieved from http://www.academia.edu/2029979/The_trending_academi

c_library_job_market_An_
analysis_of_ads_from_2011_with_comparisons_to_1996_and_
1988)

17. U.S. Department of Labor, & Bureau of Labor Statistics (2012). Librarians. In Occupational outlook handbook (2012–13) (Ed.), (Retrieved from http://www.bls.gov/ooh/education-training-and-library/librarians.htm)

18. Wang, Z., & Guarria, C. (2010). Unlocking the mystery: What academic library search committees look for in filling faculty positions. Technical Services Quarterly, 27, 66–86.

19. Wheeler, R. E., Johnson, N.P., & Manion, T. K. (2008). Choosing the top candidate: Best practices in academic law library hiring. Law Library Journal, 100, 117–135.

12

CHAPTER

How Much is Much?: A Conceptual Study of Web Traffic

Tord Høivik

Oslo and Akershus University College,

ABSTRACT

Valid and relevant statistics are required for library planning and advocacy. As libraries and their users turn to the web, library statistics must follow. In this paper we explore the use of three common traffic indicators to measure the impact of web resources from national libraries. We present and discuss the use of data on on page views, virtual visits and unique users, with examples from national, academic and public libraries. These indicators are in an early stage of development and need some conceptual and much empirical work to become good tools for strategic planning. But we note four findings: (1) the ratios between the three indicators are very unstable, so we must measure and interpret all three; (2) we find substantial differences between countries, with Denmark in a leading position; (3) in academic and national libraries the number of virtual visits is likely to overshadow the number of physical visits; (4) analysis of web traffic must be based on an understanding of J-shaped distributions ('power laws') rather than concepts drawn from ordinary well-behaved bell curves ('normal distributions').

Keywords: web analytics, national libraries, web traffic indicators, physical visits, virtual visits

INTRODUCTION

In January 2008, the British Library (BL) presented an important report on user behaviour in the virtual environment (British Library, 2008). The Director, Dame Lynne Brindley, described her institution as follows: 'We are a trusted and independent source, both in cyberspace and through our vast printed collections, with more than 67 million hits on our website in the past 12 months and 500,000 readers passing through our doors every year.' The emphasis on visits, both physical and virtual, rather than on collections or loans, was highly interesting. The numbers are large, since we speak of millions. But it is impossible to interpret them without a proper scale of measurement. The library usedhits (= page views or impressions or downloads) as the counting unit. But what do half a million physical visits and sixty-seven million page views tell us about the British Library? What do the numbers mean? And how do these indicators compare with other measures of physical and virtual traffic?

In this paper we present, discuss and compare a number of different indicators that are or could be used to measure library traffic on the web. We hope national libraries, with their large economic and intellectual resources, will take a leading role in implementing this type of usage statistics. This will allow us to understand, predict and plan the interaction between users and libraries on a more solid basis. Web statistics in libraries are still at an early stage of development. Market actors are far ahead in their use of web analytics (Wikipedia, henceforth Wp) to monitor their customers and their results. But it is time to make a start. Rather than discussing concepts and proposals for indicators in the abstract, I believe it is time to start to measure, to interpret and to argue about the meaning of empirical data gathered from the field. National libraries operate in the same environment, and compete for the same customers, as other institutions devoted to education, research, media and culture. In this article traffic data from large academic and public libraries are compared.

In the case of Norway, statistics are also compared with the most frequently visited web sites in the country.

Concepts and methods that work in the North will work equally well in the East, the West and the lovely Mediterranean South. In the future, libraries must deliver their services through three different channels: through a revitalized physical library, through branded sites on the web and, most radically, through popular web services and personalized web sites beyond the libraries proper. Norway is just one case among many. Our National Library cooperates closely with national libraries throughout Europe. We participate in the Europeana Project, which opened its virtual doors to 4.5 million digital objects at its beta launch in November 2008 (Europeana, Wp). The driving forces of change, we may name them digitalization, globalization and mass education (or professionalization), define the new rules of the game for all knowledge institutions. Most people, I am afraid, regard statistics as a trivial and boring subject best left to 'dismal scientists' like economists, demographers and statisticians. This may have been true in the past, but it will certainly not be true in the future. Digital technology provides us with vast amounts of informative data and the tools to handle them rapidly and efficiently. Globalization increases competition and forces us to compare our own countries with the rest of the world. Professionals cannot just think or talk their way to the truth. They must investigate cases and contexts and develop knowledge-based practices. The demand for good statistical data and professional statistical interpretation is increasing in all countries.

National libraries used to be the playground of scholars. The general public was served by public libraries, while students and most researchers depended on university, college and special libraries for the services they needed. With the coming of the web, national libraries will be in a very different position. The National Library of Norway plans to digitalize the totality of its holdings within the next ten to twenty years. The numbers are impressive: 4.7 million newspapers (more than 60 million pages), 4 million manuscripts, 2 million periodicals, 1.9 million pamphlets ('småtrykk'), 1.3 million pictures (photos and postcards), one million hours of radio, 450 thousand books, 250 thousand hours of movies and TV, 200 thousand music scores, 200 thousand maps, 80 thousand hours of music and 60 thousand posters (Nasjonalbiblioteket, 2007).

With regard to the general public, national libraries are likely to play a much more active role on the web than in the physical world. Digital collections can be accessed by everybody. On the web, in fact, nobody needs to know that you are a national library. Norway's national librarian, Vigdis Moe Skarstein, points out that 'we can offer knowledge, but we cannot force the user to come to us. Knowledge is increasingly something people seek and find, rather than something given to them. ... We must respond to the challenge by offering access to our quality services within the user's environment' (Nasjonalbiblioteket, 2006). In Norway, large-scale digitization is progressing at great speed. But copyright issues crop up. The law will block access to all recent documents unless these copyright issues are resolved. Norway has a tradition for collective solutions in the copyright field, and the National Library has worked hard to find ways of compensating authors for republishing their work on the web.

A large-scale experiment was approved by the authors' organizations in 2009. The project Bokhylla.no (The Digital Book Shelf) will digitalize and publish all books issued in Norway during four decades: 1990–99, 1890–99, 1790–99 and 1690–99. At the moment, nearly fifty thousand titles from the 1990s have been released, as well as nine thousand titles from earlier periods (Bang, ND). To what extent will the general public, as opposed to historians, genealogists and humanistic researchers, be interested in using these resources? Will access to this vast cultural heritage influence cultural consumption patterns? Only statistics can answer that question.

Statisticians do more than collect numbers. We compare them. The ratio between two numbers is a basic tool in applied statistics. But the numbers must be well chosen. The number of public library staff (full-time equivalents, FTEs) per ten thousand inhabitants is highly meaningful. In Norway the number lies close to four. This tells us that the average (full-time) employee serves about two thousand five hundred persons. Since only half the population are library users, the actual number of people served is about twelve hundred and fifty. The number of staff members per ten thousand is less relevant, since it lumps part-time and full-time employees together.

The number of public libraries per ten thousand does not tell us anything about library resources, however. Since libraries are municipal entities, the number (which happens to be 0.9) just expresses the ratio between the number of municipalities (about 430) and the population (five million). The number of public library branches per ten thousand, which is used in some international statistics, is not very useful either. Whether we count the number of branch libraries or the number of library organizations, we combine very small with very large units. We might as well study the importance of agriculture by counting the number of animals per farm, lumping cows, horses, pigs and poultry together.

THE BRITISH LIBRARY

Most indicators in practical library use have the mathematical form of ratios: visits per inhabitant, accessions per hundred inhabitants, loans per FTE, and so forth. When we design and discuss such indicators, we must keep in mind what the components refer to. Words point to entities. What does the numerator (visits, accessions, loans) mean? How is it measured? What entity does the denominator (population, FTE) describe? How is it measured? Finally, and most importantly: is there a meaningful conceptual relationship between the numerator, on the one hand, and the denominator, on the other? The number of staff per ten thousand inhabitants is meaningful. The number of library units per ten thousand is not.

Let me apply this thinking to the British Library and its visitors. BL is located in London, near St. Pancras station. The library serves the whole of the United Kingdom, with a population of more than sixty million. The metropolitan area of London has about thirteen million people, while the city itself has about 7.5 million inhabitants. Half a million readers a year corresponds to 6.5 visits per hundred London residents a year, four visits per hundred people living in the metropolitan area, and 0.8 visits per hundred UK inhabitants a year. These numbers are very low compared with public libraries, which typically attract a handful of visits per inhabitant per year. But such comparisons are nearly meaningless. The number of physical visitors to national libraries must be related to the purpose and intended audience of the library.

'The Library is open to everyone who has a genuine need to use its collections. ... .Historically, only those wishing to use specialised material unavailable in other public or academic libraries would be given a Reader Pass. Recently, the Library has been criticised for admitting numbers of undergraduate students, who have access to their own university libraries, to the reading rooms. The Library replied that it has always admitted undergraduates as long as they have a legitimate personal, work-related or academic research purpose.' (Wikipedia article on the British Library).

Half a million visits a year corresponds to about forty thousand visits per month, or about ten thousand visits per week, or about fifteen hundred visits per day. The primary BL audience consists of specialized researchers and scholars in the humanities and social sciences who live in London, or who are visiting the area for the purpose of research and study. But we should note that the sheer number of library visits is not of great interest as such. It is the content and duration of the visit, we could almost say: the meaning of the visit, that is important.

Visits are easy to measure by the use of turnstiles or electronic counters. But the value and impact of the library depends on what people actually do once they have passed the gates. The number of visits published by the British Library, nearly half a million in 2007, refers to the use of the reading rooms only (British Library, 2007, p. 27). The library does not, in other words, follow the IFLA guidelines, where a physical visit is defined as 'the act of a person's entering the library premises' (Poll, 2007, p. 112). Had they behaved 'properly' and counted every entry visitor, numbers would have been substantially higher. National libraries often arrange exhibitions, lectures, guided tours and other open events. In London, the BL website explains, 'the general public can see important works like the Magna Carta, Captain Cook's journal, Charlotte Brontë's Jane Eyre, ... for free in the Sir

John Ritblat Gallery.' Most national libraries would probably cast their net more widely and include all persons that enter the building, whether as scholars or as gawking tourists.

The actual design and lay-out of the building can also influence the counting. The coffee shop and the local bookstore may, for instance, be located inside or outside the library space proper. For planning purposes it is important to separate the primary users (scholars, researchers and students) from the secondary users (tourists, occasional visitors). I believe libraries ought to do rather more practical and conceptual work in this area. By a variety of simple observation methods we can, in fact, distinguish between types of use and gather detailed data on duration and content (Høivik, 2008). But such methods fall beyond the scope of this paper, which focuses on web traffic.

WEB ANALYTICS

Three indicators are widely used to measure traffic on the web: page views (PV), visits or sessions (VS) and the number of unique visitors (UV) during a given period. These are the three variables that TNS Metrix, the main provider of web analytics in Norway, publishes on their open web site. The Danish firm KPI Index, which has been engaged to measure library traffic on the web, uses the the same indicators. Wikipedia sees PV and VS as rather primitive indicators: 'Two units of measure were introduced in the mid 1990s to gauge more accurately the amount of human activity on web servers. These were page views and visits (or sessions). A page view was defined as a request made to the web server for a page, as opposed to a graphic, while a visit was defined as a sequence of requests from a uniquely identified client that expired after a certain amount of inactivity, usually 30 minutes. The page views and visits are still commonly displayed metrics, but are now considered rather rudimentary. (Wikipedia article on Web analytics). But since libraries are beginners in this field, they need to start with the rudiments of learning.

The first two indicators are additive. The number of page views, or the number of virtual visits, in weeks A and B can be added together. The number of unique visitors must be handled with care, however. UV behaves in a very different manner. The number of unique visitors in week A cannot be added to the number of unique visitors in week B in order to find the number of unique visitors in the period A+B. People who came to visit on both weeks should only be counted once. When we want to know the number ofdifferent visitors, repeat visits should be excluded. This basic fact is very often forgotten (or has never been learned). TNS Metrix measures UV on a weekly basis. The British Library used a full year.

Commercial actors have moved beyond these three and apply much more sophisticated methods to study how customers navigate their web sites. Such data are generally secret, but we can get an idea of the detail involved from Ibsen.net, which publishes detailed statistics on their web traffic (Ibsen.net, Usage statistics). The same is true in the physical world. Commercial visitor studies are an important field within marketing research. The measures that were used in the past, such as the number of physical visits, are being replaced by measures that reveal the activities undertaken by the visitors as they move around in physical (or virtual) space. The study of user behaviour is more advanced in the commercial than in the public sphere. But the relevant methods are basically the same.

PAGE VIEWS AND SESSIONS

Libraries tend to see themselves as public service providers. They live in protected environments, beyond the cut-throat competition of commercial markets. The web, however, is inherently competitive. Free or fee does not matter. All web sites compete for the scarcest commodity in the world, which is neither gold nor diamonds, but genuine human attention. Survey data from Nielsen Ratings illustrate the situation. In March 2008 the typical internet user spent thirty-three hours on the web and visited seventy different domains. He or she downloaded 1,550 web pages — and looked at each page for less than a minute. Each surfing period on the web lasted about an hour on average.

In July 2011, British Library virtual visitors spent about three minutes on the site, looking at four pages on average (Alexa, bl.uk). This is quite similar to the global Nielsen data for 2008: about forty-five seconds per page. I assume, therefore, that this was the case for BL in 2008 as well. In 2008, the web site contained about ten thousand individual pages. The average page was therefore downloaded 6,700 times a year or about twenty times a day. Sixty-seven million page views per year translate into fifty million minutes or about:

- three thousand hours of reading or attention per day
- twenty thousand hours per week
- eighty thousand hours per month
- one million hours per year.

Here I focus on scale rather than on numerical accuracy. To understand a new phenomenon, we need to know its relative size or scale, not the precise numbers. We may compare the time spent on the web with the time spent by physical visitors and by employees. Every day in 2008:

- virtual visitors spent about 400 working days at the BL web site
- physical visitors (using the seats) spent at most 1.500 working days at the London site
- staff spent 2,500 working days at the physical sites (London and Boston Spa).

So far we have considered page views. But the number of virtual visits, relative to the population served, is a more intuitive indicator. We start by defining the concept. According to ISO 2789:

- 'a virtual visit is a series of requests for data files from one and the same website visitor.

- a website visitor is either a unique and identified web browser program or an identified IP address that has accessed pages from the library's website.
- the interval between two consecutive requests must not be longer than a time-out period of 30 minutes if they are to be counted as part of the same virtual visit. A longer interval initiates a new visit.'
- Poll (2007, p. 114) adds that a virtual visit must come from
- 'outside the library premises in order to use one of the services provided by the library.
- the population to be served is the number of persons to whom the library is commissioned to provide its services.'

In 2008 the British Library did not report the number of visits. But if we assume that the number of page views per visit remained constant, 67 million page views translates into

- fifty thousand visits per day
- three hundred and fifty thousand visits per week
- 1.4 million visits per month
- 17 million visits per year.

In 2011 the BL reported that nearly four hundred thousand persons visited the reading rooms. This corresponds to eight thousand visits per week or about 1,300 visits per week day. The library has 1,200 seats, which tend to be occupied most of the day. Internet traffic was probably higher in 2011 than in 2008. Clearly, the number of virtual visits is very much higher than the number of physical visits, by a factor of forty or more. But thetime they spent at the library (web site) was substantially less. Fifty thousand virtual visits, lasting four minutes on the average, boils down to about four hundred working days. So physical duration dominates over virtual duration.

This point needs to be stressed: physical and virtual visits are very different entities. They should not be lumped together under the general heading of 'Visits'. When we work with statistics, we must compare like with like. The same principle applies to libraries and librarians: a person working 3.5 hours a week is not equivalent to a person working thirty-five hours a week. A big metropolitan library with a staff of one hundred is not equivalent to a one-person operation in a small rural community.

DENMARK

Let us look at some other libraries. In late 2007, the Danish Agency for Libraries and Media hired KPI Index to set up a standardized system for measuring web traffic. The three basic indicators (PV, VS, UV) are published every week, and are also available on a monthly and an annual basis. This is the most advanced web indicator system I am aware of in the library world. This tool provides quite a fine-grained mapping of the traffic from a national point of view. Libraries that join the system can generate a wide variety of more detailed traffic reports. To what extent such studies are actually carried out I do not know. So far the impact of these data on published library research seems modest. But KPI is definitely one of the finest data resources available for studies on web traffic.

Here I give a brief analysis of the 2010 data (Høivik, 2011). In 2010 the KPI index covered eighty-four public library systems (excluding Sydslesvig, which belongs to Germany). Their size went from half a million (central Copenhagen) to twelve thousand inhabitants. The typical size, defined as the libraries lying between the upper and the lower quartile, ranged from sixty to thirty thousand people. Half the libraries registered between 2.6 and 5.2 web visits per inhabitant in 2010. The median number was 3.6. This is somewhat below the number of physical visits, which was 5.5 in 2009. The typical number of pages downloaded per session spanned a quite narrow range, from 9.2 to 7.1 pages for the middle half. The number of visits per unique visitor also had a quite compressed distribution. The interquartile range went from 3.6 to 2.9. Bigger libraries seem to have more users than small libraries. In 2010, libraries serving more than fifty thousand had a median of seventy-nine visits per hundred inhabitants. Libraries in the 30–50 thousand range had sixty-seven. Libraries with less than thirty had fifty-nine. The differences are not major. Two likely explanations are (1) that big libraries serve more urbanized areas with more web oriented users, and (2) that big libraries offer better web sites which attract more users.

We can summarize the results as follows: the average Dane visits her public library on the web three to four times a year and consults an average of eight pages each time. Note that the number of unique users depends both on the interval (2010) and on the library system doing the counting. A roaming user will be counted as a unique visitor in several systems during the same interval. The sum of all unique users is in fact greater than the total Danish population — and belongs to the many numbers that may be computedwithout referring to a meaningful social entity.

In Danish public libraries web visits and physical visits have the same order of magnitude. The ratio between the two is therefore much lower than what we found in the British Library. To get an understanding of web traffic in general we explore this further. The Royal Danish Library serves both as a national and as a university library. It has reported on both types of visits since 2005. The virtual-to-physical ratio varied quite a bit: 6.5 in 2005, and then 8.5 - 18 - 26, and finally 11 in 2009 (Kongelige Bibliotek, 2010; Kongelige Bibliotek, 2011). Since the numbers are rough, I only use two significant digits. Some of the variation is due to changes in the actual procedures. Most libraries are still in a calibration phase. We are testing out the tools and trying to find out what to do with the numbers. So far we have found three different patterns:

- British Library: about forty virtual visits per physical visit
- Royal Danish Library: about ten virtual visits per physical visit
- Danish public libraries: less than one virtual visit per physical visit.

The double function of the Royal Danish Library probably explains the high number of physical visits. The British Library allows students, but is meant for research rather than for ordinary student use. As national libraries increase their resources and services on the web, the number of physical visits is likely to be swamped by the number of virtual visits. Virtual visits are not limited by physical space and have a much greater potential for growth. But they also take place in a much more competitive environment. People who visit physically are committing their bodies as well as their minds to the library. People who arrive through the web are fickle and insubstantial as ghosts. This is the secret of Google's search engine. More than half the web visitors to the British Library come directly from Google, and more than half go back to Google when they are finished (Alexa, bl.uk). Google does not tie people down. It just helps the ghosts to find the hosts (which they want to haunt, one might say).

The biggest public library in Norway, Deichmanske bibliotek in Oslo, is one of the few Norwegian libraries for which some web statistics are available. In recent years (2007–10) it has reported about twelve million page hits a year (Høivik, 2011). Since Oslo approaches six hundred thousand inhabitants, this corresponds to twenty downloads per inhabitant. In Denmark, the median number of downloads (2010) ranges from thirty-four for large to twenty-five pages per inhabitant for smaller libraries. So public library web traffic is clearly higher in Denmark than in Norway. It is also very much better documented. A reasonable and testable hypothesis would be: web sites with user friendly design (information architecture) and rich resources will be used more intensively than sites with less content and a less attractive structure. This implies longer dwell times and more page views per visit. On top of that, I would expect central urban libraries to receive more web traffic than more provincial and rural libraries. The digital transformation starts in the more cosmopolitan urban environments.

UNIQUE VISITORS

On the web, the location of the library is no longer important. The virtual arm of the British Library serves not only the United Kingdom, but all speakers of English as long as they have access to the web. The number of people who can read the English language may be above one thousand million (Wikipedia). About five hundred million have access to the web (Internet world stats). These are all potential users of the BL web site. But people living outside UK will, in general, have national libraries of their own. Theprimary audience of the British Library web services must be the sixty million inhabitants of the UK. We know, however, that a few percent of the incoming traffic to BL comes via the Indian version of Google. The share was 4% on July 11, 2011, Alexa says. In addition to page views, BL also provided some data on unique visitors (or hosts). The numbers lie close to five million:

- 2005/06: 4.2 million unique visitors
- 2006/07: 4.9 million

The number of unique hosts is the best approximation available of the number of individual users of the website. During 2006/07 less than ten percent of the UK population consulted the BL web site. We do not know the distribution of traffic between the United Kingdom, on the one hand, and the Anglophone world beyond the Isles, on the other. In Germany, approximately forty percent of the NL web traffic came from outside the country in 2006 (top level domain = .de) (Deutsche Nationalbibliothek). If we assume an equal division between national and international traffic in Great Britain, we get a participation rate of 2.45 (million UV) / 60, or about four percent of the population. The BL presumably has access to detailed traffic data that can give us better numbers, and much more information about the web users than the published statistics allow. Since we know the number of visitors, we can also calculate the average number of pages consulted:

- 2005/06: 11.7 pages per unique visitor
- 2006/07: 12.4.

With a viewing time of about one minute per page, we could imagine five million users spending twelve minutes each on the website during the year. This mental image should not be trusted too far, however. When we look at web traffic, we are dealing with data that do not follow the customary bell curves ('normal distributions'). Individual usage of library sites is characterized by J-shaped (or long-tailed) rather than by bell-shaped distributions. In the BL case, fifty-five million UK inhabitants do not consult the web site at all. Most of the active users are likely to be brief and occasional visitors.

In social life economists and sociologists often find distributions that follow the Pareto principle: that twenty percent of the individuals represent eighty percent of the activity (Wikipedia). We encounter many such distributions in studies of other library activities. I am willing to guess that we will find the same pattern on the web. If I am right, eighty percent of the sixty-seven million pages — or about fifty million pages, will have been downloaded by about one million 'heavy users'. This corresponds to fifty pages (or minutes) per person per year. The remaining thirteen million pages are shared among four million 'light users'. Each of them drops in for a brief visit or two, consulting, on average, three pages a year. Once we get more detailed data on unique users and other aspects of web traffic, we will be able to test whether this 'Pareto hypothesis' is correct.

COMPARING NATIONAL LIBRARIES

Until 1998, the Norwegian National Library combined the functions of a national and a university library. Today it has the same specialized profile as the British Library. Its traditional building in the centre of Oslo was nearly totally rebuilt a few years ago. This is where researchers come to study and the general public to attend lectures and to visit exhibitions. But the library also has a vast storage, development and digitalization unit located at Mo i Rana, a former factory town in Northern Norway. Here visitors are few and far between. The library registered 170 thousand physical visits in 2008 and slightly fewer in 2009. It does not provide regular reports on its web traffic, but has released monthly data on page hits from January 2006 to April 2008 (Solbakk, 2008). At that time the annual number of page hits was around eight million per year. This implies that the Norwegian level of traffic, as measured by page hits, was about fifty percent above the British level:

- United Kingdom: 67 million page views / 60 million inhabitants = 1.1 pages per inhabitant
- Norway: 8 million pages / 4.6 million inhabitants = 1.7 pages per inhabitant.

In the years 2002–07 the National Library of Finland registered between 1.0 and 1.5 page views per inhabitant, placing it between Norway and the UK. In 2010 the library had 170 thousand physical and 1.9 million virtual visits. This places it close to the Danish National library:

- Finland: 11 virtual visits per physical visit (10)
- Denmark: 11 virtual visits per physical visit (09).

Web Traffic: the Case of Norway

TNS Gallup dominates the Norwegian market for analysis of web traffic. They publish a weekly summary from more than one hundred organizations, covering the three indicators we have mentioned. The information released by the National Library only gives the number of pages. I will therefore concentrate on the page indicator — with data from Week 20, 2008 (= May 12–18) (TNS Gallup, 2008):

- two sites registered more than one hundred million page views per week: Norway's biggest newspaper (VG) and Finn, which is the biggest site for classified advertisements
- nine sites notched up between ten million and a hundred million downloads: the Norwegian MSN site, two national newspapers, one digital news service, the main national broadcasting web site, two search portals, one phone directory and one financial magazine.
- thirty-nine sites delivered more than one million, but less than ten million, page views.

- thirty-seven sites had more than one hundred thousand, and less than one million, page views.

We find the National Library web site, with a weekly average of 160 thousand hits, towards the lower end of this last category. To place the traffic in context, we may look at the site's immediate neighbours. These range between one hundred and two hundred and fifty thousand page impressions per week. Just above the National Library we find:

- one small regional newspaper (Østlandets Blad)
- one small national newspaper for farmers (Nationen)
- an important soccer support club (Rosenborg)
- a regional news site (iBergen.no)
- a very small phone directory (Folk.no)
- a magazine on technology (Teknisk Ukeblad).

Just below we encounter

- a marketing magazine (Kampanje)
- a youth-oriented magazine (Det Nye)
- a web site on buildings and construction (Bygg.no)
- consumer magazine on sound and image (Lyd & Bilde)
- a web site devoted to cell phones (Mobilen)
- a consumer technology web site (Teknofil).

These are, we may say, small-scale neighbours. The most popular web sites in Norway attract, in other words, a thousand times as much traffic as the National Library. Other major web services receive several hundred times more attention. Many regional and specialized web sites also attract much more traffic. With the ambitious and costly goal of digitizing all its collections within two decades, the Norwegian National Library has set its sight on the new, digital knowledge-based society. Involving the general public in this project is still a major task. In a country known for its many tunnels underground, roads still need a fair amount of traffic to justify their existence. Libraries that provide access without attracting users have not achieved their goals. I believe this is true in the rest of Europe as well.

CONCLUSION

Libraries of the future will clearly deliver their services in three different environments: physical buildings, local web sites and distributed user environments. The physical libraries are caught up in a process of profound change. When their customers change, they must adapt or fade. It is the new ways of research, teaching and learning that give the process its momentum. Public libraries are becoming more physical and more digital at the same time. Academic librarians are taking a new look at their place within the intellectual division of labour. Student-oriented libraries redefine themselves as learning centres. Research-oriented libraries take on much more active roles in the research process. The term cyberinfrastructure is clumsy, but covers what I mean. National libraries face similar challenges. On the web they must shift from collection-based thinking to competitive customer services. This involves a deep change in organizational culture.

The speed and uptake of web-based practices will of course differ from country to country. But I see the end result as given. The digital revolution will impose its conditions on the world like the industrial revolution did two hundred years ago. We are all moving in the same direction. When we discuss the future of national libraries, we can therefore draw on trends and experiences from many neighbouring fields. Library statistics tend to focus on information that is easy to collect but hard to interpret. The number of physical visits does not indicate the time spent at, or the benefit gained from, using the library. The number of loans does not measure reading or understanding, learning or wisdom. If we look at web-based versus physical activities, we see that virtual visits are different in kind from physical visits.

During its first decade (1991–2000) the web was mainly used to strengthen existing structures and services. Web sites were typically designed as mirror images of the organization chart. Beaujolais Nouveau was poured into old bottles. During the second decade, from 2001 to 2010, the web started to challenge the structures themselves. This third decade (2011–2020) will be a period of major readjustment and deep structural change. All the libraries I have visited in this paper are confronting the challenges of the web. Their strategic plans are clear enough: a new era is upon us. What I argue for is better documentation. We need statistics that show us what is happening. Some of the numbers will be encouraging. Others will not. Professionals should gather, publish and discuss both kinds, beyond hope or fear.

In Norway, massive digitalization of books started after this paper was first drafted. So far we have only scattered data on the corresponding traffic. But the volume is clearly increasing. In the Summer of 2010 the national librarian reported that traffic to the book site now 'represents roughly one book page per second, day and night.' (Høivik, 2010). This implies thirty million pages a year, moving the library from the low to the high end of its category (= 0.1–1 million page hits per week). However, to properly interpret these changes we need to look at all three indicators. The book readers on the web form a relatively small group. They make a big impact because they remain on-site for long periods and download lots of pages: 65 pages during the average visit. To understand web traffic, we need to separate high-intensity from low-intensity users. Since we lack data about user profiles from other national and academic libraries, this is a task for the future.

WIKIPEDIA ARTICLES

British Library, http://en.wikipedia.org/wiki/British_Library

English language, http://en.wikipedia.org/wiki/English_language

Europeana, http://en.wikipedia.org/wiki/Europeana

Pareto principle, http://en.wikipedia.org/wiki/Pareto_principle

Web analytics, http://en.wikipedia.org/wiki/Web_analytics

REFERENCES

1. Alexa, bl.uk.
2. Bang, S. (ND): Om bokhylla (About the Digital Book Shelf).
3. British Library (2007): The British Library Annual Report and Accounts 2006/07, URL= http://www.bl.uk/about/annual/2006to2007/performstats.html
4. British Library (2008): Information Behaviour of the Researcher of the Future. URL= http://www.bl.uk/news/pdf/googlegen.pdf
5. Deutsche Nationalbibliothek (2006): Report on usage in 2006. Page no longer available.
6. Høivik, T.: KPI Index Denmark
7. Høivik, T. (2008): Count The Traffic. Paper for the 2008 IFLA conference in Quebec.
8. Høivik, T. (2010): Få, men trofaste brukere. Samstat blog, July 2010.
9. Høivik, T. (2011): ST 54/11: Data fra Deichman 2010.
10. Ibsen.net, usage statistics. http://www.ibsen.net/stats/ [accessed 11 December 2011)

11. Internet World Stats. Internet world users by language.

12. Kongelige Bibliotek, Det (2010): Årsberetning 2009 (Annual report 2009). København: Det Kongelige Bibliotek.

13. Kongelige Bibliotek, Det (2011): Årsrapport 2010 (Annual report 2010). København: Det Kongelige Bibliotek.

14. KPI index. Homepage (Danish).

15. Nasjonalbiblioteket (2006): Nasjonalbibliotekarens kommentar til bibliotekutredningen (The National Librarian's response to the Library White Paper).Oslo: Nasjonalbiblioteket.

16. Nasjonalbiblioteket (2007): Digitalisering av bøker i Nasjonalbiblioteket – metodikk og erfaringer (Digitalization of books in the National Library - methods and experiences). Oslo: Nasjonalbiblioteket. PDF.

17. Poll, R. and Peter te Boekhorst (2007): *Measuring quality. Performance measurement in libraries*. München: Saur, 2nd edition.

18. Solbakk, S.A. (Director of digital services, Nasjonalbiblioteket) (2008): Personal communication, May 23-24.

19. TNS Gallup. Topplisten uke 20/2008.

Index

A

academic libraries, 1, 3, 4, 10, 213, 223, 228, 235, 244, 257

Academic libraries, 40, 44, 132, 192, 194, 218

Assessment, 14, 53, 63

C

Collections assessment, 14

Consortia, 113, 114, 115, 118, 130, 132

E

Ebooks, 113, 122, 123, 124, 131, 133, 134

e-readers, 117, 177, 178, 179, 187, 188

ethnography, 40, 57

F

Flipped classroom, 60

H

higher education, 1, 2, 16, 18, 211, 213

I

Information literacy, 14, 17, 57, 60

J

Job hunting, 218

L

library anxiety, 39, 40, 44, 45, 47, 50, 53, 56

Licensing, 113

O

Ontario Council of University, 113, 114, 118, 132, 133

outreach, 39, 40, 41, 42, 43, 44, 48, 53, 54, 138, 139

P

physical visits, 239, 240, 246, 248, 249, 250, 253, 256

T

Twitter, 135, 136, 137, 138, 151, 152, 153, 154, 155, 156

V

virtual visits, 240

W

web analytics, 240, 245
web traffic indicators, 240